Introduction to Design for Civil Engineers

R. S. Narayanan and A. W. Beeby

SPON PRESS
Taylor & Francis Group

London and New York

First published 2001 by Spon Press
11 New Fetter Lane, London EC4P 4EE

Simultaneously published in the USA and Canada
by Spon Press
29 West 35th Street, New York, NY 10001

Spon Press is an imprint of the Taylor & Francis Group

© 2001 R.S. Narayanan and A.W. Beeby

Typeset in Goudy by Bookcraft Ltd, Stroud, Gloucestershire.
Printed and bound in Great Britain by Biddles Ltd., Guildford and
King's Lynn.

British Library Cataloguing in Publication Data
A catalogue record for this book is available from the British Library

Library of Congress Cataloging in Publication Data
Introduction to design for civil engineers / R.S. Narayanan and A.W.
Beeby.
        p. cm.
    Includes bibliographical references.
    1. Engineering design. 2. Civil engineering. I. Beeby, A. W. II.
Title.
    TA174..N35 2000
    624–dc21   00–044548

ISBN 0–419–23550–7 (pb)

Contents

Preface   vii
Acknowledgements   viii

1   What is design?   1

2   Types of failure   4

Functional failure 5
Serviceability failures 5
Disconnection failures 5
Translation failures 6
Elastic instability failures 6
Material overstress failures 7

3   The elements of a structure and how they function   8

Introduction 8
Structures and members 8
Stress and strain 11
How a beam works 13
Stresses and strains in beams 27
Efficiency of sections subjected to bending 34
Columns or struts 41
Tension members 43
Continuity 43

4   Structural form and materials   55

Introduction 55
Stone 57
Bricks 60
Timber 60
Iron and steel 62

Reinforced concrete 64
Prestressed concrete 66
Discussion 68

5 Loads and loading   **70**

Introduction 70
Sources of loads and classification 70
Statistical concepts used in conjunction with loading 78
Discussion of some types of loads 80

6 Material properties   **96**

Introduction 96
Reinforced concrete 100
Structural steel 113
Masonry 118
Structural timber 125

7 Structural safety, limit state design and codes of practice   **132**

Introduction: what do we mean by safety? 132
Sources of uncertainty 133
Means of providing safety taking account of uncertainty 139
Examples of use of safety formats 149
Accidents and robustness 155
Building regulations and design codes 158

8 The design process   **173**

Introduction 173
Factors governing the engineers' task 174
Project stages 175
Concluding remarks 178

9 Designer in a changing world   **179**

Introduction 179
Continuing professional education/development (CPE/CPD) 180
Sustainability issues 182
Learning from failures 186

Index   191

Preface

Our aim in writing this book is to bring out the broad issues affecting all design. We have not attempted to provide prescriptions, formulae or recipes to solve particular problems. These can all be found in codes, manuals and other books. We have instead concentrated on the factors affecting design and the process undertaken. Engineering education tends to be a compartmentalised learning of different techniques. The main missing link is the appreciation of the purpose behind these techniques, and the context of their application. We have tried to illustrate the all-pervasive nature of design.

The book is particularly aimed at young engineers and undergraduates. We hope that we have kindled their curiosity, so that students feel that their studies are purposeful. We believe that this in turn will allow the students to derive more from their curriculum. It is hoped that the book will enable them to appreciate the bigger picture, and how many of the skills they had acquired in a disparate fashion can be brought to play in a constructive manner.

The book has been a pleasure to write, and our hope is that it contributes to the understanding of the creative activity that is design.

Acknowledgements

The authors gratefully acknowledge the valuable help received from the staff of Cadogan Tietz. In particular we wish to thank Mrs Janet Everard for patiently typing a large part of the manuscript, and Mr Richard Stone for his splendid work on the figures.

Chapter 1

What is design?

The word 'design' means different things to different people. To some, a designed object will be a slick and smart-looking object – it has aesthetic appeal. To others it will be a thing that performs everything it is called upon to do – it has functional efficiency. To yet others it may be something that represents good value for money. In truth, all the above statements are valid. A long and rigorous process has to be gone through to achieve the desired end results and that process is design. Let us illustrate this through two examples.

Imagine a town with a river flowing through it, cutting it into two. Let us further assume that the only mode of crossing the river currently available is by boat, which is highly restrictive and inconvenient to the communities. If you are asked to improve the situation, how will you go about it? You will first think of speedier alternatives to boat crossing, e.g. a bridge or a tunnel. Locating the new crossing or crossings (in design you should not close off options too early) will need many considerations – e.g. the existing network of roads on both sides, the ground conditions, the resulting length of crossing (cost implications), and so on. You might analyse the total impact of two or three alternatives.

Having chosen an ideal location, preliminary surveys (topographical, soil, traffic and so on) will be commissioned. Preliminary details will then be prepared, together with a budget. With such major construction, obtaining planning consent could be a fraught process. The scheme could affect some people adversely or will be perceived to do so by some sections of the community. The planning process provides for hearing all objections. You will, therefore, have the task of presenting the case for the scheme, highlighting such issues as economic, environmental and employment benefits. You will also need to counter the technical arguments of the objectors. After successfully negotiating the planning process, you must prepare detailed proposals for constructing the crossing. At every stage a number of options will need to be considered, e.g. materials to be used (steel, concrete, etc) and their form (number of spans, suspension, cable stayed, beam and slab). When the design is optimised, detailed drawings are prepared to enable tenders to be invited. When a contractor is appointed, you will need to supply detailed drawings and specifications so that the product you had in mind is actually constructed. This whole process of delivering to a client a product that had its genesis in the form of vague requirements is design. The details will vary from project to project.

Take another example: that of building a hospital for a Health Authority which has decided on the type (general or specialist) of the hospital and the number of patients to be catered for. Here the designer will need to consider the best location, taking into account the centres of population the hospital seeks to serve and infrastructure facilities such as road access, public transport, utilities, waste disposal, etc. Space requirements for the various departments will need to be addressed, as well as the relationship of the various spaces. Hospitals are heavily serviced and consideration must be given to heating and ventilation of spaces, supply of medical gases, emergency alarms, call bells, operating theatre requirements and so on. There will also be issues such as boiler plant location, safe storage of medical gases, disposal of medical wastes, central laundry, patient lifts, fire and security precautions, communications within and without the hospital. Externally, vehicle circulation will need careful planning with allowance for ambulances, cars, public transport, and pedestrian traffic. The buildings and surroundings will need to be as pleasant as possible.

As in the previous example, at every stage a number of alternatives will need to be considered and choices will need to be made. The load-carrying skeleton of the building, i.e. the structure, will have to be designed taking note of the constraints imposed by other requirements for the efficient functioning of the hospital. These will include the location of column supports and the effect of the circulation of services, which may need holes through the structure. As before, detailed drawings will need to be supplied for tender and then for construction. The contractor who works to the drawings will expect them to have been co-ordinated between the various disciplines, bearing in mind the particular aspects of the brief so that everything fits. That is the mark of a good design.

From these examples it can be seen that the genesis of the design process lies in some basic and simple requirement of a client, such as a river crossing or a hospital. The final form of the facility is the result of design. The impact of design does not end with construction. Ease or otherwise of operability, inspectability and maintainability, are all inherent in the chosen design and these have huge implications for the client and users.

Another feature of design that should have come across in these examples is that most design is multifaceted, requiring the skills of many disciplines, such as civil/structural engineering, services engineering and architecture. Designers thus work in teams. The lead role in the team is usually dictated by the nature of the scheme, e.g. a civil engineer for the river crossing example and an architect for the hospital.

We can summarise some attributes of a good design:

- fulfilment of all client's requirements
- functional efficiency
- value for money

- sensible balance between capital and maintenance costs
- buildability, maintainability and openability
- aesthetically pleasing.

Practical design is usually a compromise between all of these. Every client will attach different weighting to the above-noted features.

From the above, we can provide a possible definition for design.

> Design is an optimisation process of all aspects of a client's brief. It requires the integration of all the requirements to produce a whole that is efficient, economic and aesthetically acceptable.

Optimisation implies an iterative approach. Integration of various requirements demands collaborative work. Emphasis on the 'whole' requires an awareness beyond one's own specialism. Being efficient, economic and aesthetically pleasing, all at once, implies compromises and a trade-off between different requirements.

The examples would also have illustrated the steps in any project, which can be summarised as follows:

- briefing
- project investigations
- sketch designs
- planning consent
- detailed design
- working drawings
- tender
- construction/commission.

So how does one become a good designer? Here are a few tips:

- good education and training in the chosen discipline
- general awareness of the workings of other allied disciplines
- keeping up to date by following technical journals, books and participating in the activities of professional institutions and other learned bodies, generally maintaining technical curiosity throughout your life
- developing communication skills, including oral, written, drawn and electronic.

It can thus be seen that design is not all to do with the calculation of stresses and strains, nor does it have precise, fixed outcomes. It is a much more holistic, creative and satisfying pursuit.

Chapter 2

Types of failure

A major objective of design is to avoid failures; most of this book is indirectly concerned with introducing the methods used for ensuring that failures do not occur. There are, in fact, a number of different types of failure and, before entering into the study of these methods, it is essential to understand the differences between them.

We shall start by considering failure in its broadest sense and illustrate this by a story related to one of the authors by his father, who was for many years concerned with construction in various parts of the world. The supply of water to an industrial plant for which his company acted as consultant required the construction of a water tower in order to provide sufficient water pressure at the plant. The tower had to be fairly tall and was to be built of reinforced concrete. A small local contractor was given the job of building the tower. He had no mechanical equipment and the concrete was all mixed by hand and taken to the tower in baskets carried on the heads of women. The quality of the placing and compaction of the concrete was very poor and, by the time that the tower was about half built, it was clear that the tower could never be built to its full height and be filled with water. The resident engineer acting for the consultant condemned the tower and asked that it be demolished and a new tower built. The local government officials were not prepared to lose face by admitting that the local firms were incapable of constructing the tower, so political pressure was brought to bear and this led to a compromise. The tower would not be built any higher and the tank would be installed at the existing level. This was duly done and the tower was handed over by the contractor with, no doubt, great pride. It was, however, totally useless to the client because it was not high enough to do what it was required to do. The tower did not fall down; nevertheless it was a failure as far as the client was concerned as it was just as useless to him as if it had fallen down. This suggests a broad definition of failure along the following lines:

A structure has failed if it is, or becomes, incapable of fulfilling its required purpose.

Clearly not all failures are of equal seriousness. The example of the water tower above is clearly a case where the failure was total as the resulting tower was useless. However, many failures, though they inhibit the proper functioning of a structure, either do not make it completely useless or the fault can be rectified without excessive expenditure.

We can now look at various ways in which this may occur.

Functional failure

The above example comes into this category. This is where the structure is simply unsuitable for its required function. There is no failure of the materials from which the structure is built, and the structure probably meets all the requirements of the appropriate design rules and regulations. Failures of this type, or at least partial failures, are less uncommon than we might like to think. They are not necessarily the fault of the structural engineer or constructor; it is actually more likely that the fault lies in the client either not having analysed sufficiently carefully what he needed or not having communicated his requirements to the design team with sufficient clarity. Drains not placed at the lowest point of the area they are intended to drain, bridges with insufficient clearance for what is intended to pass under them, have all happened and fit into this category. The way to avoid such failures is for all parties to analyse the problem being solved and for all to be absolutely clear about what is required. This is probably the first task that should be carried out in any design.

Serviceability failures

Serviceability failures are where the behaviour of the structure under the service loads inhibits the proper functioning of the structure. Examples of this type of failure are very frequently due to excessive deflection. For example a beam that deflects to the extent that it distorts doorframes and makes it impossible to open the doors would be a serviceability failure. Another example is where vibration of the structure, possibly due to machinery, makes it impossible to operate sensitive scientific equipment. Most serviceability failures are related to a lack of stiffness of the structure rather than a lack of strength.

Disconnection failures

If you take a set of children's wooden bricks and build a tower, it eventually reaches a height where it becomes unstable and falls down. This is a failure of the tower but the materials from which the tower is built (the bricks) are completely undamaged and can immediately be used again to build another tower. This type of failure, where there is no overstress or failure of the materials but failure arises simply because the members of the structure become separated, is a *disconnection failure*. Another example of such a failure is the collapse of a house of cards. It will be seen that disconnection failures tend to be sudden and catastrophic and therefore, in the real world of structures, need to be avoided.

There are practical examples of such failures, frequently called 'progressive collapses'. Probably the most famous is Ronan Point, where a gas explosion in a multistorey residential structure made of precast panels resulted in a wall being blown out. Loss of this wall removed support from the structure above, which fell. The resulting loads from the falling panels caused the lower levels to collapse and the whole corner of the building fell down. This failure is discussed again in a later chapter. It is arguable that the collapse of masonry buildings where the masonry is laid in lime mortar are also disconnection failures as the masonry units (stone, brick) are often not failed. This type of failure probably kills more people than any other in earthquakes in less developed countries.

In large buildings, the possibility of this type of failure is nowadays guarded against by ensuring that structural members are connected together with connections capable of resisting at least some minimum force.

Translation failures

These are failures where there is no failure of the structure (at least initially). They are generally the result of a failure of the foundation or soil around a structure. The classic example of an incipient translation failure is the Leaning Tower of Pisa. The Tower itself is not failing, there is simply excessive differential settlement between one side of the tower and the other, causing it to lean. Should the tower fall down, the prime cause will be a rotational translation of the tower. Of course, as the tower falls, and certainly when it hits the ground, it will fall to pieces, but the breaking up of the tower would be a secondary issue and not the primary mechanism leading to failure. With luck, and the application of engineering expertise, the tower will not collapse but will be straightened somewhat.

Actual translation failures do occur, and are probably the normal way in which one would expect a retaining wall to fail.

Elastic instability failures

When you take a straight piece of thin wire, hold it vertically and try to apply a vertical load to the top, you will find that the load that can be applied is small, and the wire will bend. Take the load off and the wire will return to its original position, showing that the material from which the wire was made has not been overstressed. Failure of the 'structure' in this case is the result of the deflection and is generally known as 'buckling' or, more correctly, as elastic instability due to second-order effects. The mechanisms involved will be discussed further in Chapter 3. For the present, it is only necessary to see that the failure does not arise primarily from overstress of the material but is the result of the deflection, which is a function of the stiffness of the member rather than its strength. Elastic instability can lead to failure in almost any type of thin member subjected to compressive stress. Examples of cases where this possibility has to be considered are:

- the design of slender columns
- the compression zone of thin beams
- the outstanding flanges of rolled steel sections where these are in compression
- the walls of tanks, silos etc. where these are in compression
- the steel plating in box girder bridges.

Material overstress failures

Failures in this category are where the member or structure is overloaded to the extent that the material fails. Guarding against this type of failure probably forms the major part of the detailed design of a structure. Since the next chapter is largely concerned with understanding this type of failure, nothing further will be said here.

Chapter 3

The elements of a structure and how they function

Introduction

In this chapter, an attempt will be made to list and describe the function of the various elements from which structures are made. There is, inevitably, a certain amount of mathematics involved in this process but it has been kept to the minimum necessary for the understanding of the behaviour of members and hence for gaining an understanding of the effect on the performance of the variables considered.

Structures and members

This section simply sets out the names of the various elements commonly found in structures and defines their function. How they fulfil this function will be covered in later sections.

An essential starting point is to define a *structure*. Many definitions have been proposed, but these will not be presented nor their relative merits debated here. For the purposes of this book, the structure is the collection of elements within a construction that are assumed, and designed, to support the loads applied to the structure and transmit them safely to the foundations.

Structures can have many forms; they may not even be visible to the casual observer. For example, Figure 3.1 shows the steel frame for a multi-storey building under construction. The steel frame acts as the 'skeleton' of the building and, like the skeleton of a human being, when the building is complete and clad in brick, glass or stone, the skeleton will not be visible. Nevertheless it is the skeleton that supports the entire load: the outside cladding, the floors, services and so-forth being hung from or otherwise supported on the frame. In other cases, the structure is clearly visible. For example, the load-bearing elements in a normal domestic house are commonly the exterior walls. Similarly, almost all that you see of a bridge is 'structure' and is essential to supporting the loads that the bridge is designed to carry.

Let us now look at the parts, or *members*, that make up a typical structure. A framed structure will be used as the basis for this categorisation.

Figure 3.1 Steel skeleton to a building.

The primary elements in a frame are those that most immediately support the applied loads. These are generally the floor *slabs*. The function of the slabs is to transmit the loading from where it is applied to those members that support the slabs. This requires the slabs to transfer the loads in a direction perpendicular to the direction of the loading. Normally, the applied loading acts vertically, because this is the way gravity acts; the slabs have to transfer the load horizontally to supporting beams, walls or columns.

The second form of member is a *beam*. A beam collects the load from one or more slabs and transmits it to the members supporting the beam. This may be another beam, or a wall or column. Beams behave in the same way as slabs but, because they concentrate the load from slabs, they tend to carry much higher load intensities.

A particular form of a beam is a *truss*. This is a beam made up of small individual units usually arranged to form a triangulated structure.

Columns collect the loads from beams and slabs and transmit them downward to the foundations. They behave in a fundamentally different way from beams and slabs in that they mainly transmit loads in a direction parallel to the axis of the member (i.e. the column is generally vertical and the load is being transferred downwards).

Structurally *walls* carry out the same function as columns, that is, they transmit loads downwards. In tall buildings they also serve an important function in stiffening a building against lateral loads (i.e. wind). Architecturally they serve to divide up a building into compartments and to provide an outer skin. Generally walls are fairly lightly loaded. It should be noted that there are cases where walls function more like slabs than columns. An example of this is where a wall forms the vertical sides of a tank that contains liquid. Here the wall is mainly subjected to a horizontal load from the liquid and has to transfer this load vertically to supports. A wall that supports soil (a retaining wall) is similarly behaving structurally more as a slab than a wall.

Foundations take the loads from the columns and walls and transfer them to the underlying soil or rock. Because the soil is normally much weaker than the material forming the structure, the foundation generally has to spread the load over a sufficient area of the soil for the stresses in the soil to be limited to levels that will not cause excessive settlements.

A *tie* relatively rare form of member. It carries a load by tension. Members that carry only compression are frequently called *struts*, particularly if they are elements in a truss.

This concept of a hierarchy of members that transmit the load from where it is applied to the underlying soil is true for all structures, though not all members are present in all types of structure. Figure 3.2 attempts to illustrate the above discussion.

From the paragraph on walls it can be noted that the terminology is inexact and other forms of nomenclature are used. Also, particular types of structure tend to be described in more specialised jargon. For example, a column supporting a bridge is generally described as a *pier* and a wall supporting the end of a bridge as an *abutment*.

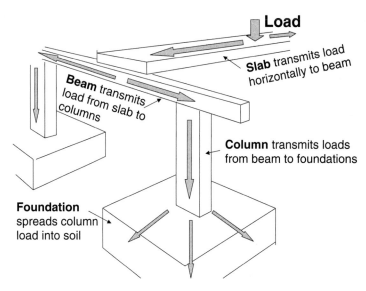

Load

Slab transmits load horizontally to beam

Beam transmits load from slab to columns

Column transmits loads from beam to foundations

Foundation spreads column load into soil

Figure 3.2 Hierarchy of members in a structure.

Stress and strain

Before embarking on an explanation of how beams, or any other parts of a structure, behave, the concepts of stress and strain need to be understood. These ideas are quite simple but are fundamental to all our thinking about how members resist load. We need some means of characterising the materials that are going to be used in a structure. This characterisation should be independent of the size and shape of member that we may wish to design. The principal description used in structural engineering is the stress–strain curve for the material. *Stress* is a convenient way of expressing load intensity; it is simply the force applied to a given area divided by the area. The units of stress are thus force per unit area, commonly N/mm^2. Intuitively, this seems a reasonable parameter to choose; if we know the strength of a member under, say, pure compression, then we expect the strength of a member with twice the cross-sectional area to be twice as strong. Assuming that the stress at which failure occurs is constant, then the force that a member will carry is the stress multiplied by the area, and a member of twice the area will be predicted to carry twice the load.

All materials deform under load. Though these deformations may well be too small to see with the naked eye, they may be measured using instruments available in any materials or structures laboratory. Suppose we measure the deformation of a short member subjected to pure compression: we will find that the member shortens slightly. To be able to apply the information we get from this test to other situations, it is convenient to express this deformation as a deformation per unit length of the specimen. This is the *strain*.

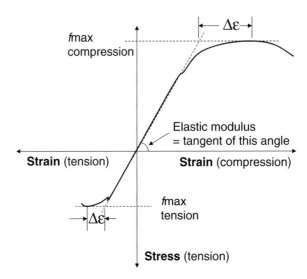

Figure 3.3 A stress–strain curve.

If a specimen is loaded axially in a testing machine, the stress can be computed from the load applied by the testing machine, and the strain can be calculated from the measured deformations. The results are then normally presented as a graph of stress plotted against strain. This is the *stress–strain* curve for the material. Such a curve is shown in Figure 3.3. There are various features of this curve that are common to all normal structural materials and are important in understanding or predicting the behaviour of structures made from the material.

The first point to note is that a major part of the curve is close to being a straight line passing through the origin (the point where stress and strain are both zero). This is a feature of all normal structural materials and is a fact used extensively in design. A material that has a straight-line stress–strain response is said to be *elastic*, and most materials are elastic over a major part of the load range that they can withstand. The properties of an elastic stress–strain curve can be defined uniquely by the slope of the line (i.e. the tangent of the angle between the strain axis and the stress–strain line). This slope is called the *elastic modulus* (occasionally the modulus of elasticity or Young's modulus). If the elastic modulus is known, then the stress–strain curve over that range of stress within which the material behaves elastically can be drawn. The elastic modulus defines the stiffness of a material; a material with a low modulus of elasticity will deform substantially more than a material with a high modulus of elasticity under a given load.

The second point to note is the ultimate stress, f_{max} that is reached by the material. This is the measure of the *strength* of the material and for structural engineers it is generally the most important property of the material.

The third point to note is that, at higher levels of stress, the curve shown in Figure 3.3 deviates from the initial straight line with the strains becoming considerably larger as the material approaches failure. The deviation from the elastic line at the maximum stress ($\Delta\varepsilon$ in Figure 3.3) is a measure of the *ductility* of the material. If $\Delta\varepsilon$ is small, or non-existent, then failure of the material is sudden, like breaking a stick of chalk, and the material is described as being *brittle*. If $\Delta\varepsilon$ is large, the material will undergo large deformations before failure and the material is described as being *ductile*. A paper clip, for example, can be bent and straightened extensively without breaking; the paper clip is made of a very ductile material. Construction materials range from brittle (stone and the new fibre reinforced composites) to ductile (structural steel). In many aspects of design, a degree of ductility is essential. For example, in earthquakes most deaths result from the collapse of old-fashioned masonry buildings that have minimal ductility and just fall down at the first shake; new buildings, properly designed to be ductile, will be damaged by earthquakes but should not actually collapse.

The part of the stress–strain curve where the stress is tensile is not necessarily similar to the part where the material is subjected to compression. Generally the elastic modulus is the same but the strength in tension and the ductility may be very different. Concrete or stone, for example, have a tensile strength that is very low compared with their compressive strength (commonly about 10%). In contrast, steel has roughly the same strength in tension as in compression and is very ductile both in tension and compression.

How a beam works

Beams and slabs are probably the commonest and most fundamental parts of a structure and it therefore seems appropriate to start by attempting to give an understanding of how they function and hence how they may be designed.

Galileo's analysis

The first person to attempt a scientific study of structures was Galileo. Galileo completed his book, A *Discourse on Two New Sciences,* in his old age in 1636 while he was under house arrest after his celebrated trial and the recantation of his views on the Copernican system. The book, published in 1638, develops an analysis of the behaviour of beams. Mostly, this analysis was correct and it presents a picture of how a beam behaves that is as correct today as when he wrote it. He did make one fundamental mistake which, considering the level of knowledge at the time, was excusable and will be explained later. What we shall do here is to follow Galileo's approach, but without making Galileo's error.

The principle of a *lever* or seesaw had been understood at least since the time of the Greek philosophers. Though the rule that a seesaw or lever will balance when the product of the force multiplied by the distance from the fulcrum on one side is equal to that on the other may well have been known empirically by much earlier people, Aristotle seems to have been the first to state it in writing. It was Archimedes, however, who gave a proof of the principle that we should accept today. Galileo would have

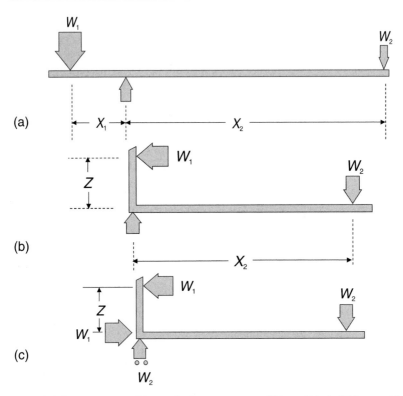

Figure 3.4 Lever systems: (a) straight lever or seesaw ($W_1x_1 = W_2x_2$); (b) lever with right-angle bend at fulcrum ($W_1z = W_2x_2$); (c) complete system of forces on cranked lever.

been well aware of Archimedes' work, as would have been the people he was address-ing in his book. We know, and Galileo knew, that, for the system shown in Figure 3.4(a) to balance, the equation $W_1x_1 = W_2x_2$ must be satisfied. The lever does not have to be a single straight bar; the same relationship holds for the lever with a right-angle bend in it shown in Figure 3.4(b). Here, because the short leg of the lever is ver-tical, we will call the dimension from the fulcrum to the force W_1 z rather than x_1. Hence, we can write, for a situation where the forces are just in balance, that $W_1z = W_2x_2$. In fact, there is a problem with the system sketched in Figure 3.4(b) and, sup-posing that there was no friction between the fulcrum and whatever supports it, the lever would move rapidly to the left out of the drawing. To retain equilibrium, there must be another horizontal force acting on the system to balance the horizontal force W_1. This is shown in Figure 3.4(c). We now have a system where the vertical forces acting on the lever balance (the load W_1 downward at the end of the lever is balanced by an equal upward force at the fulcrum) as also do the horizontal forces. As sketched in Figure 3.4(c), the system will not rotate about the fulcrum nor will it move up or down nor to the left or right; it is completely in balance.

Figure 3.5 Illustration of cantilever from Galileo's *A Discourse on Two New Sciences*, 1636.

Galileo started his analysis by considering a *cantilever*, which is a beam that is built rigidly into a support at one end and is completely free at the other end. Figure 3.5 shows the illustration from Galileo's book. Though this may not look much like an illustration from a modern textbook, the system Galileo has in mind is clear enough. What he saw was that the forces acting on the cantilever beam were just the same as the forces acting on the lever shown in Figure 3.4(c). Figure 3.6(a) tries to illustrate this: if we can imagine a pin somewhere near the bottom of the beam where it is attached to the wall, then, if the beam is not to hinge downwards, there has to be a horizontal force acting to the left somewhere near the top (Figure 3.6b). It should be clear that, exactly as in the case of the lever, W_1z must be equal to W_2x_2 for the beam to stay in position. The distance z between the compressive force and the tension force is generally referred to as the *lever arm*. It will be seen that, for the cantilever not to fall down, two internal forces must develop at the support: a force to the right near the bottom of the beam and a force acting to the left near the top of the beam. These two forces must be of equal magnitude. These forces are the resultants of stresses developing in the material of the beam. It should be evident that the action of the load pressing down and the internal force W_1 at the top of the section trying to pull the beam up against the wall is tending to stretch the top part of the beam. This is therefore in tension. Since the force at the bottom of the section is the opposite of this, it is trying to compress the bottom of the beam. Thus, to balance the load, a tension force and tensile stresses must develop near the top of the beam and a compressive force and compressive stresses must develop near the bottom of the beam at the support. The magnitude of these forces must be such that:

$$W_1z = W_2x_2.$$

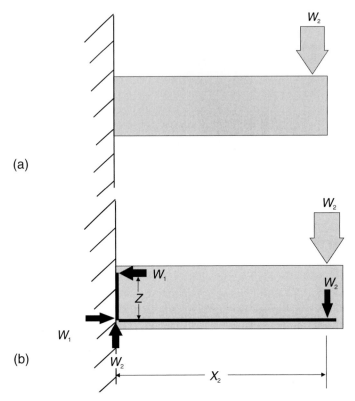

Figure 3.6 Lever analogy for cantilever: (a) cantilever; (b) force system ($W_1z = W_2x_2$).

Internal and external moments

It is convenient to consider the two sides of the equation separately; W_2x_2 is a function of the load applied to the cantilever and how far from the support we apply the load, while W_1z is a function of the forces developed within the section at the support and the depth of the section. The first of these may be considered to be external, to do with the loading on the cantilever, while the second term is internal. Put another way, we decide arbitrarily what load W_2 we wish to put on the cantilever and how far from the support we wish to put it (the distance x_2). If the cantilever is not to collapse, the horizontal compressive and tensile forces W_1 a distance z apart must develop internally at the support section to balance exactly the loading we have applied. W_2x_2 is attempting to rotate the beam in a clockwise direction about the support, and we call an action that is trying to rotate something a *moment*. We call W_2x_2 the *external moment* and W_1z the *internal moment*. (Using an alternative set of definitions, the external moment, is the *applied moment* and the internal moment is the *resisting moment*, or moment of resistance.) It will be seen that the internal moment W_1z is acting to rotate the cantilever in an anticlockwise direction about the support. We can now re-express the equation for the state of the forces on the cantilever in words by saying that the external moment must be balanced by an equal and opposite internal moment. When we design a cantilever, or any system of beams, the approach will be to establish the external moments developed by the loads we wish to put on the system and then to design a beam cross-section that is strong enough to enable the internal moment to develop.

Galileo's study now considered separately the two elements in the problem: the external moment and the internal moment. Very briefly, his study of the internal moment led to the conclusion that, at failure, the internal forces would be proportional to the cross-sectional area of the cantilever while z would be proportional to the depth of the cross-section. This gives the internal moment as being proportional to the area of the section times its depth. Thus, for a rectangular section:

$$M_r = kbh^2$$

where M_r is the internal or resisting moment, k is a constant, b is the breadth of a rectangular section, and h is the depth of a rectangular section.

Galileo was therefore able to say that, if one had two beams of different dimensions then:

$$\frac{M_{r_1}}{M_{r_2}} = \frac{b_1 h_1^2}{b_2 h_2^2}$$

where the subscripts 1 and 2 indicate the two beams.

Thus, if one beam had been tested, and was known to be able to resist a moment of M_{r_1}, then the strength of any other rectangular beam could be calculated.

A simply supported beam

A much more common form of beam than the cantilever is a *simply supported* beam, as sketched in Figure 3.7(a), and Galileo wished to establish rules for the capacity of these beams. He found that he could extend his lever analogy. Figure 3.7(b) illustrates how a simply supported beam may be considered to be two levers back-to-back and inverted relative to Figure 3.6(b). It will be seen that the internal forces at the centre of the beam from the two levers counterbalance each other. Since the levers have been inverted, it will be seen that the tensile force now acts at the bottom of the section and the compressive force at the top. By equilibrium and symmetry, it can be seen that, if a load W is applied at the middle of the beam then the upward forces developed at each support must be $W/2$. The length of the levers will be seen to be each $L/2$ and hence the external moment exerted by each lever is $W/2 \times L/2 = WL/4$. This is the moment that must be

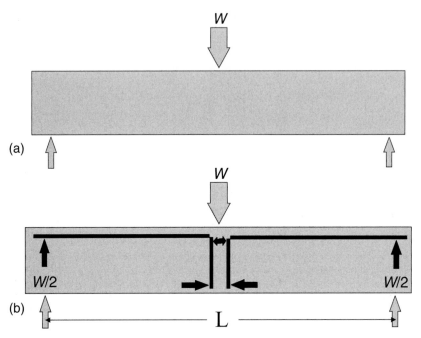

Figure 3.7 A simply supported beam: (a) beam; (b) lever analogy.

developed in the section at mid-span by the forces W_1 separated by the distance z. Thus, for a simply supported beam with a central point load:

$$M_r = M$$
$$W_1 z = WL/4$$

Thus, Galileo could say that, if he knew that a cantilever of a given cross-section and length L_c could carry a load W, then the same load W_c could be carried by a simply supported beam with the same section of length $4L_c$.

Galileo took his analysis somewhat further and considered such matters as beams with sections of different shapes (e.g. comparing the strength of a circular section with a rectangular one), but we shall not pursue this further here. On this point his answers were in error owing to the mistake he made and which will now be explained. Before this, however, it will be useful to summarise what the analysis above has achieved.

Galileo's analysis shows, by the lever analogy, how loads on a beam generate an external moment which, if the beam is not to fail, must be exactly balanced by an internal moment. This internal moment arises from a tension force developing near one face of the beam, and an equal and opposite compression force developing near the other face. The external moment is a function of the size and position of the applied load, while the internal moment will be a function of the strength of the material and the area and depth of the section.

Where Galileo was wrong

It must be emphasised that the discussion above is correct and, though it follows Galileo's line of argument, it has made use of later knowledge that had not been grasped by Galileo. The critical piece of later knowledge was first formally stated by Newton in his Third Law which states that, for static equilibrium, action and reaction must be equal and opposite. Newton was born in the year Galileo died, so there was a considerable time gap before the science of statics developed further. Newton's *Philosophiae Naturalis Principia Mathematica,* which set out his new world picture was actually published in 1686, 50 years after Galileo completed his book. It seems odd that Newton's third law had not been formulated earlier, since it seems to us fairly obvious, while the apparently less obvious concept of equilibrium of moments had been understood and formed the fundamental basis of Galileo's beam analysis (i.e. the concept of a balancing seesaw).

Galileo would seem to have been influenced in his thinking by seeing stone columns failing as beams. Pictures illustrating this appear in his book. It is possible to understand Galileo's conclusion by breaking a piece of chalk by bending it and looking at the failure. Galileo concluded that failure was due entirely to tension and that the whole section must have been subjected to tension at the moment of failure. Looking at our broken piece of chalk, this seems reasonable: there is no sign of crushing. As a result, Galileo assumed that the 'hinge' of his lever system was the bottom edge of the section, and that the whole section was in tension at failure. The necessity for a balancing compressive force at the hinge did not seem to occur to him. The result of this misapprehension was that Galileo assumed that the tension force in his cantilever acted at the centroid of the section (a uniform tensile stress acting on a section is equivalent to a tension force equal to the stress times the area of the section acting at the *centroid* of the section). The lever arm z was thus $h/2$ for a rectangular section and, if the tensile strength of the material was f then the tension force was given by bhf and hence:

$$M_r = bh^2f/2$$

We know now that the compressive force cannot act at the very bottom of the section; this would imply infinite compressive stress in the material. To be able to develop a compressive force equal to the tension force, some considerable area of the material at the bottom of the cantilever in Figure 3.6(b) must be in compression. The tension force must thus be significantly less than Galileo assumed, and the lever arm is not necessarily $h/2$. If the material used in the beam has equal strength in tension and compression, then one might conclude that the area required to develop the compressive force would be the same as that required to develop the necessary tensile force. This would suggest that half the section is in tension and half in compression rather than all in tension as assumed by Galileo. The reason why Galileo saw no signs of crushing in his columns is that stone (and chalk) is very much stronger in compression than it is in tension and his beams failed when the tensile capacity of the material was reached. This is not true of all materials.

At the time when he developed his theories, Galileo would have had no simple way of establishing the complete truth of what he was proposing. The relationships between the strengths of rectangular sections of different sizes that are given above were derived by Galileo and are correct. So are the relationships he produced between simply supported beams and cantilevers. If, therefore, tests were carried out on a single beam with one span, Galileo's relationships would correctly predict the strength of beams with other spans and section sizes. The concept of stress had not been developed nor were any means available to Galileo that would have permitted him to obtain explicit values for the ultimate compressive or tensile strength of his materials. Nor, had he thought of the idea, did any means exist to measure strains in the type of materials used for construction. His conclusion that the tensile force acted at the centroid of the section was thus untestable in Galileo's time and, indeed, for very many years later. Those of his conclusions that were testable at that time were true. Only when Newton had proposed his laws and Hooke had devised the theory of elasticity could the analysis of beams be developed further. In fact, the formulae we use now for the calculation of stresses in a beam made of elastic materials were not clearly presented until 1773 by Coulomb. Parent had developed a basically correct approach in 1713, but his published papers were rather obscure and were forgotten.

Bending moment diagrams

Galileo's analysis can be developed slightly further to provide more information about the external moments. We shall consider again the cantilever beam analysed by Galileo and shown in Figure 3.6. If we consider a section a distance x from the load then the load provides a leverage, or external moment, at this point of Wx in a clockwise direction. There must be an equal and opposite internal moment at this section if the cantilever is not to break at this point (Figure 3.8). Clearly, the external moment varies with distance from the load and, as a consequence, so must the internal moment.

We can conveniently draw a graph of the variation of the moment along the cantilever (Figure 3.9). In this figure, the moment at any distance x from the load is given by Wx and hence there is a straight-line relationship between moment and distance from the load. This graph is known as the *bending moment diagram* for the beam (often abbreviated to BMD). The bending moment diagram has been calculated from the externally applied loads and therefore is a diagram of external moments. However, since, at any section, the internal moment must be equal to the external moment, it also shows how the internal moment, and hence the internal forces, must vary along the beam. The diagram could therefore be used to

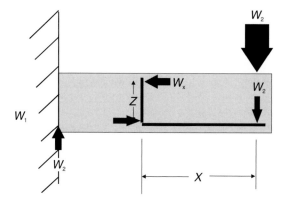

Figure 3.8 Bending moment at a distance x from the load on a cantilever ($W_x z = W_2 x$).

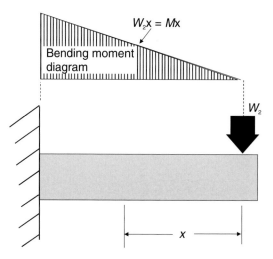

Figure 3.9 Bending moment diagram for a cantilever with a point load at the end.

establish how strong the beam needs to be at each point along its length, and we could save material by making the strength at each section only just sufficient to support this moment.

The analysis above for the bending moment diagram for a cantilever may be extended to a simply supported beam since this can be considered as two cantilevers inverted and fixed back-to-back (see Figure 3.7b). It should be evident from the analysis of the simply supported beam with a central point load that, since the reaction at the support is $W/2$ then the moment at any section between the support and the load is given by $Wx/2$, where x is the distance from the support. Since, by symmetry, the moment at a distance x from the left-hand support must be equal to the moment a distance x from the right-hand support, we can immediately draw the BMD for the beam as shown in Figure 3.10. Because we have inverted the

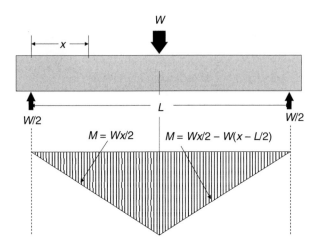

Figure 3.10 Bending moment diagram for a simply supported beam with a central point load.

cantilever, we also invert the bending moment diagram, which is now plotted downwards. In fact, there is an almost universal convention used when plotting bending moment diagrams, that bending moments are plotted perpendicularly away from the face of the beam that is in tension. In the cantilever, the forces in the top face are tensile and so the diagram is plotted upwards. In the simply supported beam, the tensile forces are on the bottom of the beam and hence the diagram is plotted downwards.

It is not intended to give a full treatment of the calculation of bending moment diagrams in this book, as that is more appropriate for a book on structural analysis. One more example will, however, be considered since it is relevant to matters that will be discussed later in this section. This is the derivation of the bending moment diagram for a simply supported beam supporting a load uniformly distributed over its whole length (Figure 3.11). This is of practical importance since, in the design of beams, the most commonly assumed type of loading is a uniformly distributed load. As with the beam with the central point load, we can immediately conclude that the reaction at the supports must be the same for both and must be equal to half the applied load. The total load is equal to the load per unit length w multiplied by the length of the beam L. Now consider a section a distance x from the left-hand support. The moment (or leverage) applied to the beam at the section considered by the left-hand reaction is $wLx/2$. However, in this case, this leverage is counterbalanced to some degree by the load on that part of the beam to the left of x. The moment resulting from a distributed load is the same as the moment resulting from a point load equal to the total distributed load on the length, considered to be acting at the centroid of the load. In this case the load on the length x of the beam to the left of x is wx and its centroid is a distance $x/2$ from the section considered. The counteracting moment is thus $wx^2/2$. The total moment at the

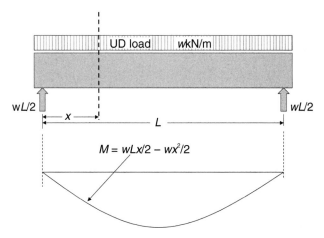

Figure 3.11 Bending moment diagram for a simply supported beam with a uniform load.

section x is thus given by the clockwise moment due to the reaction *minus* the anticlockwise moment due to the load on the length x. Thus:

$$M = wLx/2 - wx^2/2$$

Following the convention for plotting bending moment diagrams, the loading will produce tensile forces in the bottom of the beam, and hence the diagram will be plotted downwards. The moments calculated from the above equation are plotted in Figure 3.11 to give the bending moment diagram; note that the curve is parabolic in shape.

In the derivation of the equation for the bending moment for a uniformly distributed load, we considered the moments generated by the forces acting on the part of the beam to the left of the section, at a distance x from the right-hand support. What if moments had been taken of the forces acting on the beam to the right of the section considered? Let us try this.

In this case, the reaction at the right-hand end generates an anti-clockwise moment of $w(L - x)/2$ while the load on the beam gives a clockwise moment of $w(L-x)^2/2$. Assuming clockwise moments are positive, this gives the moment as:

$$M = w(L-x)^2/2 - wL(L-x)/2$$

Simplifying this equation gives:

$$M = -wLx/2 + wx^2/2$$

This will be seen to be the negative of the previous equation. It will be found always to be true that, at any section, the moment to the right of the section is

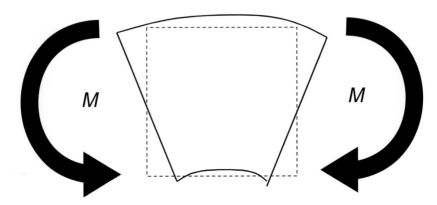

Figure 3.12 Short element of beam subjected to moment.

equal to the moment to the left of the section multiplied by −1. This is actually a consequence of Newton's third law that action and reaction must be equal and opposite. A short length of beam subjected to a moment is shown in Figure 3.12 which shows how these two equal and opposite moments bend the beam to give tension on one surface and compression on the other.

Shear forces and shear force diagrams.

In considering bending at a section a distance x from the support of a beam, we have ensured that there is equilibrium of the axial forces in the beam (i.e. the tension force on one side of the beam must be exactly equal to the compression force near the other side). We have also found that there will be equilibrium of moments (the external moment at any section caused by the loads must be exactly balanced by the internal moments resulting from the internal forces and the lever arm separating them). We have also seen that the external moment on one side of a section is exactly balanced by an equal and opposite moment from the other side. There is one further aspect of equilibrium that we have not considered: the equilibrium of the vertical forces at the section.

Figure 3.13(a) shows the part of the uniformly loaded beam considered in Figure 3.11. This section is subjected to a force acting upwards at the support (the support reaction) equal to $wL/2$ and a downward force from the load on the length x of the beam equal to wx. It will be seen that there is a net upward force of ($wL/2 − wx$) (assuming x is less than half of L). This must be balanced by an internal force acting up the section at x. This force is called the *shear force* which, for a uniformly distributed load on a simply supported beam, is given by:

$$V = wL/2 - wx$$

where V = the shear force.

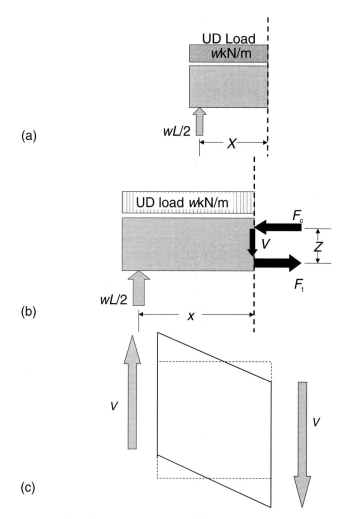

Figure 3.13 Shear forces and stresses: (a) part of uniformly loaded beam up to a section a distance *x* from the support; (b) forces acting on a short element of a beam (vertical equilibrium: $wL/2 - wx = V$); (c) deformations of a short element of beam subjected to shear forces.

If, as we did for moments, we consider the vertical equilibrium of the length of the beam to the right of the section instead of to the left, it will be found that the shear force is given by:

$$V = -wL/2 + wx$$

As for moments, if V is calculated considering loads on the length of the beam to the right of the section, it is found to be in the opposite direction to the shear

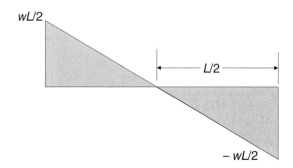

Figure 3.14 Shear force diagram for a uniformly loaded simply supported beam ($V = wL/2 - wx$).

calculated by considering the loads on the length of the beam to the left of the section. Figure 3.13(b) shows a small element of the beam subjected to these shear forces. These forces will be seen to be acting to deform the beam element as shown in Figure 3.13(c). The deformations actually lead to one diagonal of the element of beam being stretched while the other diagonal is compressed. The shear force is thus resisted by a diagonal tension in one direction and a diagonal compression at right angles to this within the beam. The material of which the beam is made must clearly be capable of supporting these stresses, as well as the horizontal tensile and compressive stresses generated by the moments.

The magnitude of the shear forces varies over the length of a beam and we have derived an equation above that will permit the shear at any section to be calculated. This equation can be used to produce a graph of the variation of shear force along the length of a beam. This is shown for the uniformly loaded, simply supported beam in Figure 3.14. This picture of the variation of the shear forces along a beam is known as a *shear force diagram*. For the particular case of the uniformly loaded beam, it will be seen from Figures 3.11 and 3.14 that the shear forces, and hence the diagonal stresses generated within the section, are greatest near the supports while the horizontal forces generated by the bending moments are greatest near the centre of the beam.

The shear force and bending moment diagrams are the fundamental pieces of information that a designer needs, to be able to design a beam since they provide the information from which the designer can calculate the internal forces and stresses at all points within the beam. These permit the selection of suitable dimensions for the beam at various sections, and the required strengths of the materials.

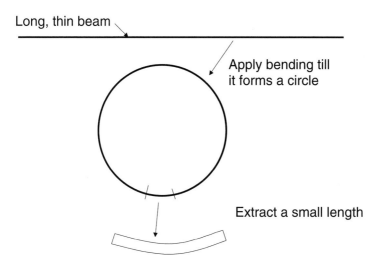

Figure 3.15 A very long, shallow beam bent into a circle.

Stresses and strains in beams

The distribution of strain over the depth of a section

In order to take our understanding of the behaviour of beams further, it is now necessary to investigate how a beam deforms when it is bent.

A very simple experiment acts as a useful starting point for this investigation. The experiment requires that we have a very long, thin 'beam' – a length of piano wire will do very well. If both ends of the piece of wire are held and flexed, the wire can be bent into a circle. Either fix the ends together or, at least, imagine them fixed together and our beam will have been converted to a perfect circle (Figure 3.15).

We are now going to consider a small length of this beam, as shown in Figure 3.16. The ends of this segment were vertical before bending while after bending the ends have been rotated to form parts of radii of the circle. The segment of the beam, when bent, subtends an angle θ at the centre of the circle and the circle has a radius r.

The analysis of the forces in a beam given above shows that, when a beam is bent, a compressive force develops on one side of the beam while an equal tensile force must develop on the other. The compressive force on one face of the beam must cause the material to shorten while the tensile force must cause the material on the other side to lengthen. Since the section we have considered is symmetrical and the tensile and compressive forces are equal, we can say that the lengthening of the tension edge of the section must be the same as the shortening of the most compressed edge and that the mid-height of the beam section will stay the same as before bending. From this, and because we know that the bent beam forms part of a circle, we

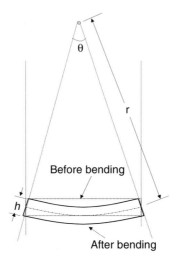

Figure 3.16 A small section of the beam shown in Figure 3.15.

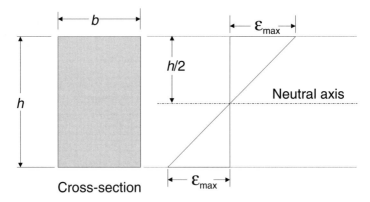

Figure 3.17 Distribution of strain over the depth of a section.

can say that the length of the top surface will be $(r - h/2)\theta$ while the length of the bottom face is $(r + h/2)\theta$. The length of the centreline of the beam is $r\theta$) and hence the extension of the bottom face is $(r + h/2)\theta - r\theta = \theta h/2$. The strain is equal to the extension divided by the length and hence the strain on the bottom face is given by $\theta h/2r\theta = (h/2)(1/r)$. The strain at the top surface will be numerically equal, but is a shortening (compressive) strain rather than a stretching (tensile) one. A little thought will show that the strain at any level within the section will be given by $(1/r)y$ where y is the distance from the centre of the section to the level considered. This distribution of strain is shown in Figure 3.17.

   The vital fact arising from this discussion is that the distribution of strain over the depth of a section subjected to bending is linear (it follows a straight line). There is some point within the section where the strain is zero, this is called the

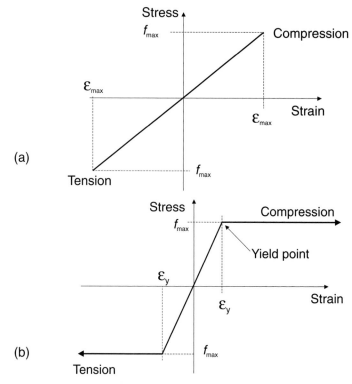

Figure 3.18 Stress–strain curves: (a) for a brittle material; (b) for an elastic-plastic material.

neutral axis. In a symmetrical section, such as is considered here, the neutral axis will be at mid-height of the section.

Derivation of the stresses from the strains

Now that the form of the distribution of the strain over the height of a beam is known, it is possible to establish the distribution of the stresses within the section and hence the location of the centroid of the tensile and compressive forces. To do this we need to know the stress–strain response of the material from which the beam is made. The stress–strain behaviour of common materials has been covered in an earlier chapter. For the purposes of this study, two idealised types of stress–strain curve will be considered: an elastic-brittle material (Figure 3.18a) and an elastic-plastic material (Figure 3.18b). Most real materials can be approximated to one or other of these types of curve, though a few, such as concrete, have intermediate forms of stress–strain curve.

The type of curve shown in Figure 3.18(a) will be considered first. It will be seen that, at the moment when the beam fails, the stress at the outermost fibres of the section will be f_{max} and the strain will be ε_{max}. At mid-height, the strain will be

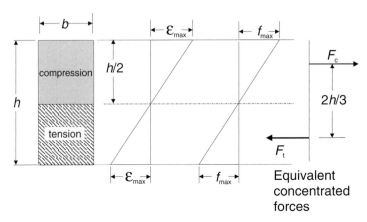

Figure 3.19 Distribution of stresses over the height of a beam made of an elastic material.

zero and, where there is zero strain, the stress–strain curve will show that there must be zero stress. If we consider a point half way between the neutral axis and the outside edge then, from the strain diagram in Figure 3.17 it will be seen that the strain is $\varepsilon_{max}/2$ and hence, from the stress–strain diagram, the stress will be $f_{max}/2$. It should be evident that, because the stress is directly proportional to the strain, the linear distribution of strain leads to a linear distribution of stress (Figure 3.19).

In order to find the moment that the section can support, it is necessary to know the magnitude of the tensile and compressive forces and the centroid of the force. Just to clarify this point, since we have a distribution of stress over the section, it should be noted that a distributed stress has the same effect as a single concentrated force acting at the centroid of the distributed forces, and we wish to work with these equivalent concentrated forces. To do this we must establish their magnitude and where they act.

The general equation for establishing the total compressive force is:

$$F_c = \int b_y \sigma_y dy$$

where: F_c is the compressive force, b_y is the breadth of the section at a height y from the neutral axis and σ_y is the stress at a height y from the neutral axis. The integration is carried out for the range of values of y from the neutral axis ($y = 0$) to the compression face of the member ($y = h/2$).

In fact, for the rectangular section we are considering, the force can be calculated by inspection. If the stress is uniform over the whole of the upper part of the section, then the force equals the stress multiplied by the area over which it acts. Where the stress varies, the force is equal to the area over which the stresses act multiplied by the average stress. Since the distribution of the stress in the compressed part of the section is triangular, the average stress is half the maximum stress. So the force is given by half the area of the section (the area over which compressive stresses act) multiplied by half the maximum stress (the average stress). This gives:

$$F_c = bhf_{max}/4$$

The tension force need not be calculated, since we know it must be equal to the compressive force.

While an equation could be written to calculate the position of the centroid of the compressive force, this, too, can be found by inspection for this simple case. The distribution of stress is triangular, and the centroid of a triangle is 1/3 of the distance from its base. Since the height of the triangle is $h/2$, we can immediately say that the equivalent concentrated load acts at a distance $h/6$ from the top face of the section. Similarly, the tension force must act at a height of $h/6$ from the bottom of the section.

The lever arm (the distance between the compressive and tension forces – see Galileo's analysis of a beam above) can be seen to be $h - h/6 - h/6 = 2h/3$. The internal moment is now given by the force multiplied by the lever arm and thus:

$$M = bhf_{max}/4 \times 2h/3 = bh^2f_{max}/6$$

This is the *elastic moment of resistance* of a rectangular section. Though it will not be derived here, a general equation for this that will work for any shape of section is:

$$M = f_{max}I/Y$$

where I is the second moment of area of the section (sometimes called the *moment of inertia*) which for a rectangular section is given by $bh^3/12$), and Y is the distance from the neutral axis to the outermost edge of the section which, for a rectangular section, is $h/2$.

Substituting the appropriate values for a rectangular section into the general equation will give the answer above.

We shall now carry out the same calculations for the slightly more complicated case of the elastic-plastic stress–strain curve. Here we shall look at two cases: first the case where the maximum strain is twice the yield strain $2\varepsilon_y$, and second where the strain is very large compared with ε_y.

Elastic-plastic stress–strain curve: maximum strain = $2\varepsilon_y$

Having defined the maximum strain, the distribution of stresses can be established from the stress–strain diagram.

At the top of the section, where the strain is $2\varepsilon_y$, the stress–strain curve shows the stress to be f_{max}. If a point mid-way between the neutral axis and the top face is considered, then, from the strain diagram, the strain is half the strain at the top,

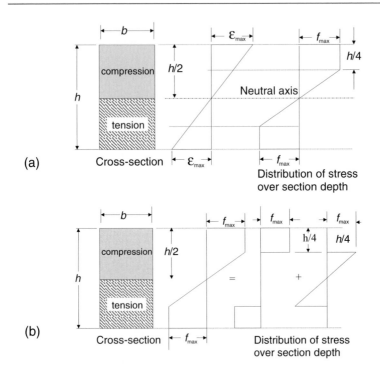

Figure 3.20 Distribution of stresses in a beam made from an elastic-plastic material: (a) stress distribution; (b) breakdown of stress distribution for ease of calculation.

which is ε_y. At this strain, the stress will again be found to be f_{max}. If a point one quarter of the height of the compression zone from the neutral axis is considered, then the strain will be $\varepsilon_y/2$ and, from the stress –strain diagram, the stress will be $f_{max}/2$. These values will be seen to lead to the distribution of stress shown in Figure 3.20(a).

The easiest way to calculate the moment is to break down the stress distributions into two forces in compression and two equal and opposite forces in tension. Thus we have a compressive and tensile force acting near the top and bottom of the section due to the constant stress acting on the upper and lower quarters of the section, and a compressive and tensile force acting closer to the neutral axis due to the triangular stress distributions acting on the inner quarters of the section (Figure 3.20b).

The upper compressive force may be seen by inspection to be equal to $(bh/4)f_{max}$, and it will act at a distance $h/8$ from the top of the section. The lower compressive force will be equal to half the upper force and will act $(h/4 + h/12) = h/3$ from the top of the section. The tension forces will be the same and will act at the same distances from the bottom of the section.

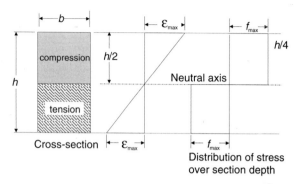

Figure 3.21 Distribution of stresses in a beam made from a plastic material.

The moment will now be given by:

$$M = (bh/4)f_{max} \times 3h/4 + (bh/8)f_{max} \times 2(h/2-h/3) = 0.229bh^2f_{max}$$

Elastic–plastic stress–strain curve: maximum strain very large

As the maximum strain increases, the area of the section subjected to a triangular stress distribution gets smaller until, at very large strains, the triangular part can be ignored and it can be assumed that there is a uniform stress of f_{max} in compression over the top half of the section and the same in tension over the bottom half (Figure 3.21).

The compression force is now $bhf_{max}/2$ and the lever arm is $h/2$, resulting in a moment of $bh^2f_{max}/4$. This is the *plastic moment of resistance* of the section.

Similar calculations could be carried out for other maximum strains and a graph of moment against maximum strain plotted (Figure 3.22). If the maximum strain is equal to the yield strain, then the equation for the moment will be that for the elastic moment of resistance. Below this, the moment will be directly proportional to the maximum strain. It will be seen that the plastic moment of resistance is, in this case, 50% higher than the elastic moment of resistance and that the moment is very close to the plastic moment of resistance for maximum strains greater than about four times the yield strain.

The elastic moment of resistance and the plastic moment of resistance provide useful limiting cases and, in practice, much design work assumes either one or the other. For example, the elastic formula will give a good estimate of the strength of brittle materials such as glass, ceramics or plastics. Furthermore, most structural materials are approximately elastic under loads that are well below failure and, if we wish to limit the stresses present under normal service or operating conditions, then the elastic equation can again be used. If we are considering the ultimate strength of a mild steel bar, however, the plastic moment of resistance will be the most appropriate. Some materials cannot be considered to be fully plastic at

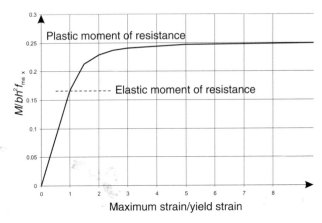

Figure 3.22 Development of moment as a function of maximum strain for a beam made from an elastic-plastic material.

failure, but neither are they entirely brittle. For example the equations developed for calculating the moment of resistance of reinforced concrete members assume a behaviour somewhere between these two limiting cases.

Efficiency of sections subjected to bending

Improving the efficiency by changing the height–depth ratio of a rectangular section

It is interesting to gain a picture of the size of the forces that are developed internally in a beam compared with the forces that are actually being carried by the beam. It has been shown in Galileo's analysis of a beam that the maximum external moment produced in a simply supported beam with a load applied at the centre is given by $WL/4$. This moment must be resisted by an internal moment that is equal to the tensile or compressive force multiplied by the lever arm z. Thus:

$$WL/4 = f_c z$$

If the material of the beam can be considered to be elastic, then it has been shown that the lever arm is equal to $2h/3$. If we substitute this value into the above equation and rearrange, we get:

$$F_c = F_t = 0.375WL/h$$

If a beam has a span that is 20 times its depth (a reasonable practical situation) then it will be seen that the internal forces (F_c and F_t) are 7.5 times the applied load W. The internal forces developed in a section thus tend to be much larger than the external applied loads. Beams thus have to be made of materials that are very strong, both in tension and compression. The shallower the beam relative to its span, the stronger the materials need to be.

We can, however, attempt to arrange the material in a section to give us the most effective shape so that we maximise the load that can be carried by a beam made of a particular material. A look at the equations derived above for the moment of resistance of a beam may help here. The equations for the elastic and plastic moment of resistance both have the form:

$$M = kbh^2 f_{max}$$

For a given material, therefore, the moment of resistance increases proportionally with the section breadth b and proportionally with the square of the depth. If we consider a constant amount of material then the volume of the beam bhL must remain constant and thus, for a given span, bh must be constant. As bh is the area of the cross-section A:

$$M = kAhf_{max}$$
$$= Kh$$

where kAf_{max} is K, which is constant in this case. It will be seen from this that it will be most economic to make h as large as possible, even though this will result in b becoming small since bh must remain constant.

Improving economy by changing the form of the section

There are further ways in which the economy of the section can be improved. If the shape of the section were changed so that as much of the material as possible was concentrated close to the bottom and the top of the section, the material would mostly be concentrated in the area where the strains were greatest, and hence the stresses and forces in the beam would be increased. In the rectangular section we have been considering, the material in the section close to the neutral axis is subjected to very low stress and thus adds very little to the total compressive or tensile force. Furthermore, if material were moved towards the outside edges of the beam section, the lever arm z would be increased. The three sections drawn in Figure 3.23 all have the same cross-sectional area and therefore use the same amount of material, but the I section has the material concentrated where it will do the most good. The elastic moments of resistance of the sections are in the ratio 1:2:6.5. The I section is a form of beam very commonly used in structural steel construction and the logic for its use should now be clear.

Yet further economy can be made. The vertical part of the I section (known as the *web*; the horizontal parts at top and bottom are known as *flanges*) is fairly lowly stressed. The web is important; its major function is to carry the shear forces, but the

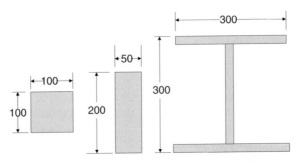

Figure 3.23 Various sections having the same cross-sectional area. The ratios of the elastic moments of resistance are 1:2:6.5

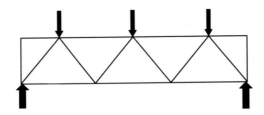

Figure 3.24 A truss.

area required for this is commonly fairly small. Much of this material can be removed. Since, as noted earlier, shear forces develop diagonal compressive and tensile stresses, the most efficient form of the structure between the flanges for supporting the shears is a series of inclined elements and, if all the material not needed is cut away, we are left with the structure shown in Figure 3.24. This is a *truss* and it provides a very lightweight way to support loads across a span. Trusses are not normally cut out of a solid piece of material, but are fabricated from smaller elements bolted, welded or nailed together (depending on the material being used).

There is a further means by which material can be removed from a beam and thus further improve the efficiency and economy of a structure. It will be remembered from the chapter on bending moment diagrams that the bending moment is at a maximum only at the critical section. Elsewhere the moment is lower and therefore, assuming that the beam is of constant cross-section, the material is not stressed to its maximum allowable value. One way of taking advantage of this is to make the beam shallower as the moment reduces. At each section, the depth must be chosen so that the forces in the section remain the same but the lever arm is reduced so that the moment (force × lever arm) remains just that required to balance the effect of the applied loads. To do this, the depth of the section has to be made proportional to the bending moment at each section. Figure 3.25 shows this for a beam supporting a uniformly distributed load. It will be remembered that a uniformly distributed load results in a parabolic bending moment diagram, and thus the depth of our beam varies parabolically. Theoretically, there is no moment at the support, so the depth of the beam on this basis should be zero. Clearly, this is

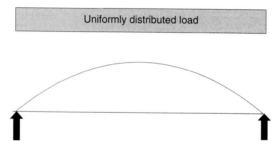

Figure 3.25 A beam with a depth proportional to the bending moment.

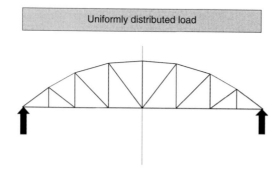

Figure 3.26 A truss of varying depth.

impossible, so the beam has been given some depth at the support. In practice, this depth could be chosen so that it is sufficient to support the shear which, it will be remembered, is a maximum at the support.

As with the uniform section beam, it is possible to remove material from the centre of the section and concentrate it on the top and bottom faces of the beam. Following the same steps as before, this will result in a variable-depth truss (Figure 3.26).

We can, however, go further in this case. At any point in the curved top member of the truss, there is a compressive force acting along the direction of the member. The horizontal component of this force just balances the tension force in the bottom member and, when multiplied by the lever arm, just balances the external moment at that section. There is also a vertical component of the force in the top member and it turns out that this vertical force just balances the vertical force or shear from the applied loads. This can be demonstrated mathematically by considering the triangle of forces shown in Figure 3.27. The slope of the top of the beam at the point considered is θ. By simple trigonometry, it can be seen that the vertical component of the force in the top member of the truss is $H\tan\theta$. The depth of the section has been chosen so that H is constant over the whole length of the beam and, at any section x, the resistance moment given by Hz_x is equal to the external moment. The gradient of the top, $\tan\theta$, is given by the differential of z with respect to x (i.e. $\tan\theta = dz_x/dx$) hence the vertical component of the force $V = H\, dz_x/dx$. We know that the shear force at any

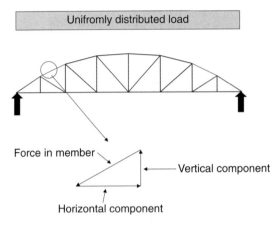

Figure 3.27 Triangle of forces for the top member of a truss of depth varying proportionally to the bending moment.

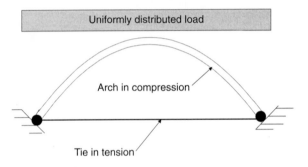

Figure 3.28 A bowstring arch.

section can be found from the differential of the moment, thus the shear force = $dM/dx = H\, dz_x/dx$. This is the vertical component of the force in the top member of the truss. Hence the top member of the truss effectively carries all the shear, and the truss members between the compression and tension members of the truss are not needed to carry any force and can be removed.

The resulting structure, shown in Figure 3.28, is known as a *bowstring arch*. The curved upper member carries the compressive forces and the horizontal 'bowstring' carries the tension.

Providing one of the forces externally

The horizontal tension can be replaced by an external horizontal force applied at the supports if the supports are strong enough to sustain it. If this is done, the result is an arch. The arch itself is entirely in compression and the balancing force is developed as a horizontal reaction from the abutments (Figure 3.29). Arches are particularly useful where masonry is to be used to span a gap. Masonry consists of

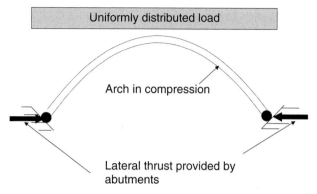

Figure 3.29 An arch.

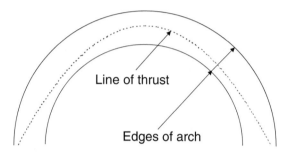

Figure 3.30 Line of thrust in a circular arch.

stone or brick units set in mortar. The mortar is weak and does not primarily act as a 'glue' sticking the units together but rather acts as a bedding. As a result, masonry has very little tensile strength, but remains very strong in compression. Masonry cannot, therefore be used to make beams but it can be used to make arches since these are entirely in compression. It is not essential for the shape of the arch to be exactly the same as the shape of a bending moment diagram for a simply supported beam. What happens is that the centroid of the compressive force in the arch must follow the shape of the bending moment diagram. Provided that the centroid of the compressive force lies well within the structure of the arch at all points, the arch will function satisfactorily. The line defining the position of the centroid of the compressive force at all points within the arch is known as the *line of thrust*.

Figure 3.30 illustrates how the line of thrust, even though it may be parabolic, can still lie within the structure of a semicircular arch ring. Semicircular arches are probably the commonest forms of arch. Up until about 150 years ago, masonry was the principal material for building heavy-duty long-span structures such as bridges, and there are still some 9000 masonry arch bridges in service in the UK (Figure 3.31). Masonry arches have excellent durability and will last almost

Figure 3.31 A masonry arch bridge.

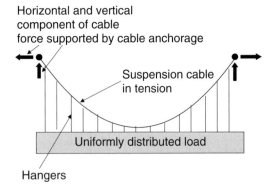

Figure 3.32 A suspension bridge.

indefinitely provided that the abutments are strong enough to resist the horizontal forces needed to balance the compression in the arch ring.

A final variation can be developed on this theme of how to span a gap most economically. To do this, it must be remembered that the development of the above argument leading to the arch assumed that the part of the beam carrying tension was left horizontal, and the parts of the beam carrying the compression were curved to the shape of the bending moment diagram. Letting the tension member take up the shape of the bending moment diagram, we could equally well have developed the argument. Had we done this, the final step in the argument would have left us with a structure that was entirely in tension. This is the basis of the suspension bridge which, in terms of weight of materials, is probably the most efficient way of spanning a gap (Figure 3.32).

Columns or struts

We have now dealt fairly thoroughly with the behaviour of beams, and slabs may be considered to behave in the same basic manner. So we now need to look at the other main components of structures; these are the members that carry loads vertically or, more properly, parallel to their axes rather than perpendicular to their axes. These members are generally dominantly in compression so, in this section, we'll look at members subject to compressive loading. Such a member is shown schematically in Figure 3.33.

Assuming that the load is applied exactly on the centroid of the section, all sections of the strut will be subjected to a uniform stress that is equal to the applied load divided by the cross-sectional area of the member. Thus, if f_{max} is the maximum stress that the material from which the strut is made can withstand, then the strut will be able to support a load equal to f_{max} multiplied by the cross-sectional area of the strut. For struts where the length is fairly short compared with the cross-sectional dimensions, this is approximately true. Unfortunately, it does not remain true for more slender struts.

What actually occurs may be established fairly easily by some simple experiments. The simplest way to explore this is possibly to take an easily obtainable, and not too strong material, such as spaghetti and test different lengths of it to establish a relationship between length and strength. This can be done quite simply by resting the bottom of the stick of spaghetti on a set of electronic scales and pressing on the top until failure takes place. If the strength of the spaghetti struts is plotted against their length, a relationship such as that shown in Figure 3.34 should be obtained. Mathematically, it can be shown that the strength should be proportional to $1/(\text{length})^2$.

Why does this reduction in strength occur with increase in length? Observation of how the spaghetti struts behave as they are loaded provides the clue. It will be noticed that it is impossible to load the spaghetti without the spaghetti deflecting, even though the loading may appear to be close to axial. There are two reasons why the element deflects: first, it is impossible to apply a load exactly on the centroid of the section and, second, the material from which the spaghetti is made is not absolutely homogeneous nor will the spaghetti be perfectly straight. The effect of these imperfections is to introduce into the strut a moment that causes some deflection. Once a deflection occurs, a larger moment is developed that is proportional to the deflection. This may be seen from Figure 3.35 where, by taking moments at mid-height of the strut, it may be observed that the mid point is subjected to a moment equal to the applied load multiplied by the deflection at mid height. The strut is thus subjected not only to an axial load but also to an axial load plus a moment. This moment will reduce the load-carrying capacity. The

Figure 3.33 A strut or column.

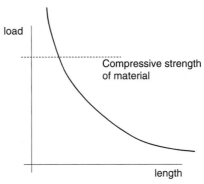

Figure 3.34 Load capacity as a function of length of strut.

flexibility of the strut, and hence the deflection and the moment, increases with increasing length, and this leads to the result shown in Figure 3.34.

The classical analysis of an axially loaded strut with pinned ends was carried out by Euler, who developed the following formula for the strength of a strut:

$$P_e = \pi^2 EI/L^2$$

where: P_e is the maximum load that the strut can sustain (often called the *Euler load*), E is the elastic modulus of the material from which the strut is made, I is the second moment of area of the section of the strut, and L the length of the strut.

It is interesting to note that the Euler load is independent of the strength of the material and is only dependent on its stiffness. At very short lengths, the Euler formula will give a load capacity in excess of the material strength multiplied by the cross-sectional area. This is clearly unreasonable, and the material strength multiplied by the cross-sectional area must give a maximum, with the Euler formula applying only where it gives a strength less than this.

Further work by later researchers, notably Rankine, led to the formulation of practical design formulae that could take account of eccentric loading at the ends of the strut and different end conditions. Furthermore, the classical analyses assume that the material from which the struts are made is elastic-brittle. Where

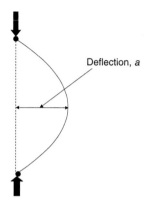

Figure 3.35 Deflection of a slender strut under load.

this is not true, rather more sophisticated analyses are necessary. These developments have not, however, changed the basic conclusion that deflections will always occur and that these deflections will increase with increasing length, leading to a reduced strength.

Fortunately, most columns in practical structures tend to be sufficiently short for the effects of deflection to be small, often small enough to be completely ignored. This may not, however, be true of members in some trusses.

Tension members

These can be dealt with very easily; they do not suffer from problems of deflection and so may always be designed to carry a load equal to their cross-sectional area multiplied by the maximum allowable tensile stress. This makes tension members very economical for long elements and is the reason why suspension structures, where the main load-carrying members are in tension, use so little material compared with other means of carrying load.

Continuity

General

So far in this chapter we have considered the elements making up a structure as separate individual members. In fact, the elements are generally connected together. One reason for this is to reduce the risk of *disconnection* failures ('house of cards' failures). Another reason is that continuous structures, where the members are connected together, are more efficient at carrying load. This last point needs to be explained and that is the purpose of this section. The problem of calculating the moments and shears in a structure where the members are connected together is much more complicated than the assessment of moments and shears in single members. Most of the content of undergraduate courses in structural analy-sis is concerned with means of solving this problem, and it is not possible to develop the necessary mathematics here. All that will be attempted here is to try to give a general background to the problem. No maths will be included.

Connecting together the spans of a beam

For simplicity, we shall consider the effect of connecting together the spans of a beam. Figure 3.36 shows a structure made up of a large number of beams, with the bending moment diagram for these beams calculated on the assumption that they are not connected together in any way. The loading on each beam is assumed to be uniformly distributed, and the bending moment diagrams can be calculated using the formulae developed earlier in this chapter.

We wish to consider the difference that would be made to these bending moment diagrams if the beams, instead of being a series of unconnected spans, were actually made of a single, very long, piece of material. Figure 3.37(a) shows an enlarged picture of the situation over a support. It will be seen that the ends of the beams are not vertical: the deflection of the beams under load has caused them to rotate. In order to connect the spans together to make up a single, long beam, the ends of the spans need to be twisted so that the ends of adjacent spans are parallel. This is achieved by applying an equal and opposite moment to the ends of the adjacent spans, as in Figure 3.37(b). The ends of the beams will now fit together. If the spans were all made of one continuous piece of material, the sec-tions an infinitesimal distance away from the support on either side of the support must be parallel, so there must be an internal moment at the support section equal to Ms shown in Figure 3.37(b). A similar condition must obtain at all other supports.

To take the argument further, we need to introduce a classical theorem of structural analysis. This is the *superposition theorem*, which states that, if you wish to find the moments and forces in a structure under the action of a number of externally applied moments and forces, this can be done by calculating the moments and forces in the structure under each separate external load or moment in turn and adding the results together. No proof of this will be offered here, but readers can easily demonstrate it for themselves by taking a simply supported beam, calculating its moment when subjected to a combination of loads (using the methods discussed earlier), and comparing this with a moment diagram obtained by calculating the diagram for each load in turn and summing the results.

If moments are applied to the ends of a simply supported beam, the result is as shown in the bending moment diagram as in Figure 3.38(a). The reult of a uni-formly distributed load is shown in a bending moment diagram as in Figure 3.38(b). As a consequence of the theorem of superposition, the bending moment diagram for a span of a beam where the ends have been connected to the adjacent spans can be found by adding these two bending moment diagrams together. The moments applied at the supports are acting to bend the beam upward (this is

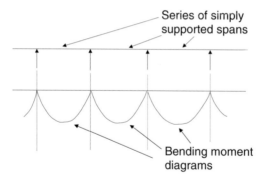

Figure 3.36 Structure made up of multiple simply supported spans.

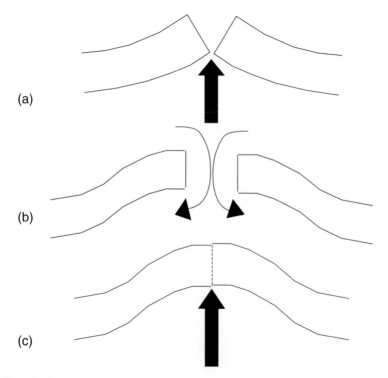

Figure 3.37 (a) Support section of beams in Figure 3.36 enlarged; (b) moments required to make ends parallel; (c) continuous connection.

known as a *hogging moment*). This is in the opposite direction to the moment caused by the loads, which tends to bend the beam downwards (a *sagging moment*). Algebraically, the moments thus have opposite signs, and summing the effect of the moments applied at the support and the loads gives the bending moment diagram shown in Figure 3.38(c). The parts of the beam near the supports are

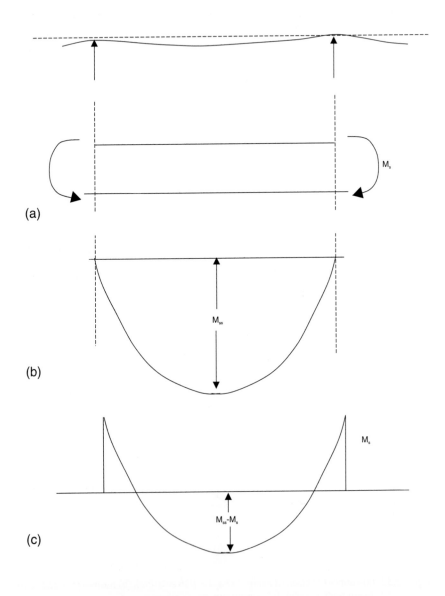

(a)

(b)

(c)

M_s

M_{ss}

M_s

$M_{ss}-M_s$

Figure 3.38 (a) Bending moment required to make support connection; (b) bending moment on a simply supported span; (c) combined bending moment.

subjected to hogging bending which generates tension forces in the top of the section and compression at the bottom, while the parts near the centre are subjected to sagging bending which generates compression at the top and tension at the bottom.

The simply supported spans were subjected to a maximum moment of M_{ss}. From Figure 3.38(c), it will be seen that the maximum moment is now either M_s over the supports or $(M_{ss} - M_s)$ at mid-span, whichever is the greater. Both must be less than M_{ss}. The maximum moment in a continuous beam is thus significantly smaller than the maximum moment in a simply supported beam. This means that either a beam of given size and material strength (giving a given moment of resistance) can carry more load if it is continuous than if it is simply supported or, for a given load, less material need be used. Continuous beams are thus more efficient than simply supported beams at carrying load. Equally important to note is that the deflection of a continuous beam will be substantially smaller than if the spans were simply supported because the added support moments tend to bend the beam upwards.

Difficulties with use of equilibrium

Nothing has been said yet about how to calculate the support moments that will occur in a continuous beam. It is here that difficulties arise. In the example discussed above, it has been assumed that the moments applied at the left-hand and right-hand support are the same. In practice this is not necessarily so. If, for example, the span attached beyond the left-hand end were different from that attached to the right-hand end, then the moment required to make the ends of the two beams parallel at the left-hand end would be different from that at the right-hand end. A more general case is shown on Figure 3.39 where the end moments differ. The moments in simply supported beams can be calculated by considering equilibrium of forces and equilibrium of moments. The general approach used, for example in calculating the bending moment diagram for a beam was as follows:

1   Calculate the reaction at one end of the beam by taking moments about the other end.
2   Calculate the other reaction by considering equilibrium of forces (sum of the reactions must be equal to the total load on the beam).
3   Calculate the moment at any section by taking moments about that section of all forces to one side of the section.

This procedure may be attempted on the beam shown in Figure 3.39.
    Start by taking moments about the right-hand support to find the left-hand reaction.

$$M_{sR} = LR_L - wL^2/2 \qquad\qquad \text{(a)}$$

This equation has two unknowns, R_L and M_{sR} so cannot be solved. In the case of the simply supported beam, we know that the moment at the support is zero, so there is only one unknown and the equation can be solved.

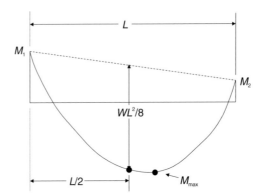

Figure 3.39 General case where support moments at the ends differ.

The process can be continued to the next stage: equilibrium of vertical forces:

$$wL = R_L + R_R \qquad \text{(b)}$$

We can substitute for R_L from the first equation to give:

$$R_R = wL - (M_{sR} + wL^2/2)/L$$

We still have two unknowns so cannot obtain a solution.

Continuing to the third stage and taking moments at a section a distance x from the left-hand support gives:

$$M_x = R_L x - wx^2/2 - M_{sR} x/L - M_{sL}(L - x)/L \quad \text{(c)}$$

Considering equations (a), (b) and (c), which correspond to the three stages in the procedure, we now have three equations and four unknowns: (R_L, R_H, M_{sR} and M_{sL}).

Quite simply, the procedure used for the simply supported beam will not work; we shall always have more unknowns than equations. As a consequence, some further consideration will have to be introduced in order to obtain more equations if a solution is to be found.

In fact, inspection of Figure 3.37 will suggest what such equations must include. The requirement in Figure 3.37 is that equal and opposite moments are applied to the ends of the two beams so that the ends finish up parallel. A suffi-cient moment therefore has to be applied to twist each end through a defined angle. The size of this angle and the moment required to twist the end through this angle, depend on the *rotational stiffness* of the beams (rotational stiffness is the moment required to twist the beam through a unit angle). This is an entirely new factor that has not been considered in any of the previous equations. In the case of continuous beams, calculation of the moments and forces must take into account the stiffness of the structure.

For convenience, we divide structures into two classes:

- *statically determinate structures*, where the distribution of moments and forces can be established by applying only the equations of equilibrium, and
- *statically indeterminate structures* where it is necessary to take account of the stiffnesses of the elements in the structure in order to establish the distribution of moments and forces.

Thus, a simply supported beam is a statically determinate structure, while a continuous beam is a statically indeterminate structure. In practice, because of the economies resulting from the provision of continuity and the extra robustness of continuous structures, most structures are designed to be continuous and therefore statically indeterminate.

Actual behaviour of continuous structures

There is an alternative approach to continuous structures from that set out above. To understand this, an experiment will be considered where the behaviour of two beams is compared. One of the beams is made of glass, which is very brittle, and the other is made of ductile carbon steel. The dimensions of the beams are chosen so that, when tested as simply supported beams over the same span, they will carry the same ultimate load. This is the same as saying that their ultimate moments of resistance are identical. The mid-span moment–deflection curves for the two beams are shown in Figure 3.40. The glass beam, for which the moment–deflection curve is shown in Figure 3.40(a), has an elastic-brittle moment–deflection diagram. At any load up to failure, the deflection is directly proportional to the applied load

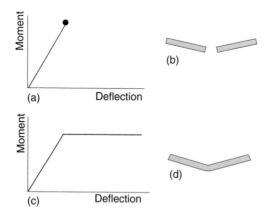

Figure 3.40 (a) Moment–deflection response of simply supported glass beam; (b) mode of failure of glass beam; (c) moment–deflection response of simply supported steel beam; (d) mode of failure of steel beam.

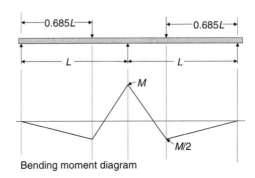

Bending moment diagram

Figure 3.41 Two span continuous beam.

or moment. If the beam were to be unloaded at any stage prior to failure, it would be found that the deflection returned exactly to zero under zero load. Failure is sudden and the beam simply falls into two parts (Figure 3.40(b)). The behaviour of the steel beam is elastic-plastic (Figure 3.40(c)). Up to the ultimate moment the behaviour is exactly as for the glass beam. Beyond the point where the failure moment is first reached, the beam will continue to deflect under a constant moment. If a small reduction is made to the load, the beam will continue to be able to support this. If the load is completely removed, it will be found that there is a considerable residual deflection. Indeed, if the beam is removed from the test rig, it will be found to have been bent permanently at the point under the load so that the beam appears as shown in Figure 3.40(d). The beam has been bent through an angle θ_p, known as the *plastic rotation*.

In a second set of tests, the beams are continuous over two spans, as shown in Figure 3.41. Under this arrangement of loads, analysis taking account of the stiffness will show that the moment over the support will be twice the maximum moment in the spans. Now consider the behaviour of the glass beam as it is loaded. As the load is increased, the moments under the loads and over the supports increase proportionally until the moment over the support reaches the ultimate moment capacity of the section. At this load the glass over the support snaps suddenly. This converts the beam into two simply supported beams but, since the load on each at this stage is 18% greater than the capacity of the simply supported beam, they both instantly fail. Figure 3.42(a) shows the load plotted against the maximum deflection for the glass beam. It will be found that the maximum deflection under any load below the failure load for the single span glass beam is 1/3 that of the simply supported glass beam under the same load.

Figure 3.42(a) also shows the load–deflection curve for the steel beam while Figure 3.42(b) shows the load–moment diagram for the section over the support, and the section at the load point. Up to the load at which the glass beam failed, the behaviour of the steel beam is the same in that the deflection is proportional to the load (i.e. it behaves elastically). At this load the support moment has reached its maximum capacity. However, instead of collapsing, the support section can

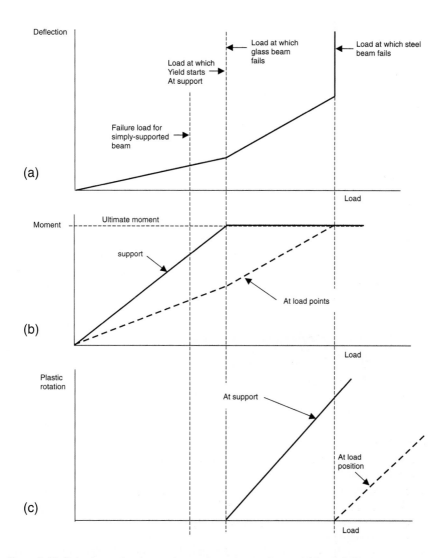

Figure 3.42 Behaviour of a glass and a steel continuous beam: (a) load-deflection curve; (b) load–moment diagram; (c) plastic rotation.

continue to carry this moment. In this case, the capacity of the section under the load has not been exceeded, and so further loading results in a constant moment at the support section while the moment at the section under the load increases more rapidly. This can continue until the section below the load reaches the ultimate moment. This will occur at a load that is 69% greater than the load carried by the simply supported beam. If the beam is taken out of the test rig at this point

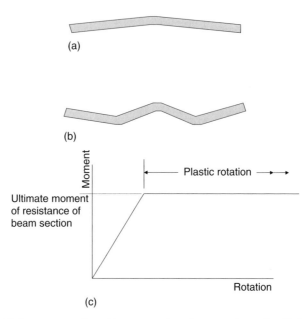

Figure 3.43 (a) Appearance of steel beam at point where maximum load is just reached; (b) appearance after further deflection under maximum load; (c) moment–rotation characteristic of beam section.

and inspected carefully, it will be found to have been bent as shown (exaggerated) in Figure 3.43(a). Plastic rotation has occurred at the support. Figure 3.43(c) shows how the plastic rotation would be expected to increase with increase in load. After the maximum load is reached, and if the load is left in place, the beam will continue to bend but this time plastic rotation will occur at both the support section and below the load points. After substantial further deformation, if the beam is removed from the test rig, it will appear as shown in Figure 3.43(b).

This experiment shows us a number of facts:

1   Continuous beams will carry more load than simply supported beams with the same bending strength.
2   Continuous beams made of ductile material such as steel will carry more load than continuous beams made of brittle materials such as glass.
3   Plastic rotations occur at critical sections before the full load capacity of the beam is reached.
4   After reaching the maximum load, the beam will continue to be able to carry this load, but the deflection will increase and plastic rotations will occur at both the support section and under the loads.
5   The deflections of continuous beams under loads below that resulting in yield at any section will be much lower that those of a simply supported beam of the same section.

A particularly important conclusion that can be drawn from this experiment is that ductility confers major advantages. The most obvious from the test is the increased load capacity but there are others, one of which is that the ability of a ductile member to deflect substantially while still supporting its maximum load makes it obvious to observers that the member is severely overstressed. This gives time either to evacuate the structure or to take action such as introduction of propping and then effect repairs.

Analysis methods

If the strengths of the critical sections are known (in the test, the ultimate moment capacities of the support section and the section at the load points) then the load capacity of the beam can be calculated. In the calculation method, the critical sections where yield will occur at failure are known as *plastic hinges* since they rotate rather like very stiff hinges. When sufficient hinges have formed for the structure to become a mechanism and to continue to deflect under constant load, then the energy put into the system by the load moving downwards by a defined amount must be equal to the energy required to rotate the hinges by the amounts corresponding to the deflection. This consideration can be used to calculate the load capacity of the system. The resulting approach to establishing the distribution of moments and forces in a structure is known as the *plastic analysis* method. It will be seen to apply specifically to the situation where a structure is just on the point of failure, and its validity depends on the material from which the structure is made being ductile. More specifically, the plastic hinges must be able to undergo the plastic rotation necessary to ensure that all plastic hinges can form.

Plastic analysis depends on knowledge of the strength of the critical sections. This is totally different from the form of analysis considered earlier in this section, where the distribution of moments depended on the stiffness of the members making up the structure. In using the stiffness approach, the stiffness of the members is commonly assumed to be constant and the materials to be elastic. It is therefore generally referred to as *elastic analysis*. We thus have two approaches to establishing the distribution of moments and forces which make the same limiting assumptions about material properties as in elastic and plastic methods for predicting section behaviour, as described earlier. Considering the experiments described earlier, glass is an elastic-brittle material and only the elastic analysis method would give the correct distribution of moments and load capacity of the glass continuous beam. The steel beam is more complex. It can be seen from Figure 3.42 that the distribution of moments in the beam up until the initiation of yield at the support in the steel beam is the same as in the glass beam. The elastic method clearly works up to this point. At failure of the beam, however, the plastic method provides the answer. In the experiments, materials that were used came closest of any to the assumptions implicit in the two analysis methods. Glass is elastic up to failure and has little or no ductility. A steel bar is close to the perfect elastic-plastic material. Other materials are not necessarily completely elastic, though most approximate to elastic behaviour at lower stress levels, nor do they have horizontal, plastic sections to their stress–strain diagrams; their behaviour lies somewhere between elastic and plastic. Though it is becoming possible to use computers to model the true behaviour of complex materials, we are still not at the stage where such approaches are practical, or necessary, for normal, everyday design. The normal procedure is to decide which approach is the more appropriate for the particular material being used and accept that the results will be somewhat approximate. Allowance for this inevitable approximation has to be made when selecting safety factors for a particular material (see Chapter 7).

In summary

- *Elastic analysis* will be used for elastic-brittle materials or where ductility cannot be guaranteed, and to establish the behaviour of most structures under levels of load corresponding to the expected service condition rather than near failure.
- *Plastic analysis* will be appropriate where the material has sufficient ductility to ensure that all necessary plastic hinges will form. In some cases this can be assumed but, in others, checks on rotation capacity are required.

Chapter 4

Structural form and materials

Introduction

The relationship between structural form and material properties is complex, but unavoidable. The relationship for a particular material must be understood by designers if they are to produce good, economical designs. It will not be possible in this book to explore this relationship very deeply: it is a subject for a book in itself. What we can do is to attempt to explore some facets of this interaction by considering some examples.

As a first example, we shall consider a case where there is only a single available material to satisfy very stringent structural and functional requirements. This example will take us to the arctic winter, where the Inuit traditionally follow a nomadic existence. Temperatures are many degrees below zero, combined with wind and blizzard. In these circumstances, shelter is required from the elements in a structure that can be built rapidly with readily-available materials. The only such material is snow and the structure that has resulted from generations of experience is the igloo. This is actually a very sophisticated structure. Typically it is made of blocks of snow about 400–500 mm thick. These are cut and laid in a spiral form until the classical hemispherical dome has been completed. The structure is self-supporting at all stages of the construction, so needs no temporary supports of any kind. The resulting form is excellent for resisting wind forces and is highly insulated, allowing the body heat of the occupants, supplemented by a fairly small blubber stove, to maintain an inside temperature some 40°C above the outside temperature. The construction has to take account of the structural properties of snow, which has some compressive strength but negligible tensile strength. The dome is ideal for this, as it is a compression structure. It also takes advantage of another property of snow: its excellent thermal insulation. The form of structure developed for this very exacting circumstance is thus the result of a combination of the function that the structure is required to serve, and the properties of the available material.

We can now move on to another apparently primitive construction material, though it is one that has been very extensively used in the past and still is in many parts of the world. This material is mud, which can be used for building in many ways. It can be made into mud bricks by being compacted into moulds and then left to dry in the sun, it can be cast like concrete into formwork, or it can be plastered onto some type of supporting material such as wattle. Mud bricks were used extensively by the ancient Egyptians and, most particularly, by the various civilisations in early Mesopotamia, an area where there was no stone and little timber. Mud bricks are not particularly strong but, if used for thick walls, can be used to build structures of considerable height. The walls of Babylon, built by Nebuchadnezzar in the sixth century BC, for example, were some 10 metres high and were said to be broad enough for three chariots to drive abreast. The main disadvantage of mud as a building material is its susceptibility to damage by water. This is probably why its major use has been in relatively arid climates. Nevertheless, it can be used in wetter climates if properly protected. Mud was, for example, once used fairly extensively in this country for the construction of cottages and some can still be found standing. Their success depended on having a good overhang to the roof, so that rain was thrown well clear of the walls, and a coating of protective limewash over the outside. The mud brickwork used in Mesopotamia tended to be protected either by fired and glazed brick on the outer face or by an imported stone facing. It is a feature of all masonry that it has minimal tensile strength and that it therefore cannot be used for beams, where, as has been seen in the last chapter, large tensions must develop. In the absence of timber to span large gaps, the arch developed.

Another aspect of the successful use of mud is in the construction of dwellings in arid regions. Here the thick mud walls of houses, usually limewashed to give a light, sun-reflecting surface, provide a means of controlling the environment within the dwelling. A problem with arid regions is the very large temperature changes that occur during the course of the day, often being very cold at night and very hot during the middle of the day. The air temperature within a building is governed largely by the temperature of the inner face of the walls. With mud brick dwellings, such as the traditional houses of the Pueblo Indians, the walls take a very long time to heat up during the day and a long time to cool down at night. As a consequence, the temperature of the inner parts of the walls stays relatively constant and this maintains the interior temperature at a relatively constant and comfortable level.

Mud, therefore, can be used successfully to fulfill most of the construction needs of a society; it can be used for the construction of massive public works such as fortifications, temples or palaces, but is also a highly versatile material for domestic use. All these applications have to recognise the properties of mud in the forms of structure that are developed: mud has reasonable compressive strength, permitting the construction of walls, platforms and arches, but is susceptible to the effects of water, requiring the surfaces to be protected. High thermal capacity confers advantages in environment control. A major advantage of mud is its ready availability and its ease of use.

This chapter has started with the consideration of two materials, mud and snow, which most readers probably would not have considered as structural materials at all yet, if used properly, even these unconsidered materials can produce highly successful structures. The lesson is that almost any material can be used structurally if its properties are properly understood and are used to develop appropriate forms of structure. We shall now move on to consider more commonly recognised, structural materials.

Stone

In many areas where people live there is an abundance of stone and this is an obvious construction material. It is not, however, without disadvantages. While forming mud into bricks is an easy process, needing little specialised equipment or expertise, stone will often require quarrying and cutting and, frequently, transporting some distance from a suitable outcrop to the construction site. It is thus generally a more expensive building material. This is not always the case, of course, and there are situations where good building stone has been readily available in an easily-worked form. An example of this is the construction of Great Zimbabwe. Figure 4.1 shows the quality of the dry stone construction in these mysterious ruins. It is clearly superb. The builders were fortunate that the area has frequent granite outcrops. This granite weathers by sloughing off thin sheets of stone that can easily be broken into suitably-sized slabs of uniform thickness for building. In general, however, except for the poorest quality of rubble masonry, stone is a substantially more expensive building material. To lay stone effectively, it is necessary to bed the stones in some form of mortar. Historically, various materials were used

Figure 4.1 Dry stone masonry at Great Zimbabwe.

Figure 4.2 Pont du Gard – a Roman aqueduct showing semicircular arches.

for this: gypsum plaster in ancient Egypt, natural bitumen in Mesopotamia and, most commonly, lime mortar. Nowadays, the mortar will normally be made using a mixture of lime and Portland Cement as this sets much more quickly than lime mortar. The mortar should not be seen as a 'glue' sticking the stones together but rather as a *bedding*. The structural properties of masonry may be considered to be the same as those of mud brick, but much better. Masonry is very strong in compression, but has minimal tensile strength. It has the great advantage over mud of being very durable.

As civilisations developed, so did skills in masonry construction, and stone masonry became the material of choice for prestige construction. The lack of tensile strength means that the most immediately obvious use for masonry is for the construction of walls and columns. The problem of using the material to span gaps was solved by the development of the arch. We have already mentioned that these were used with mud brick construction, but it is really with stone masonry that the arch came into its own and the most prolific early developers of arched structures were the Romans. The semicircular arch is probably the most characteristic feature of Roman architecture (Figure 4.2).

Two other developments used extensively by the Romans, though they were not the originators, were the vault, where an arch is extended in breadth to cover a large area, and the dome, where an arch is rotated about a vertical axis through mid-span. Domes were possibly most highly developed in the eastern part of the Roman Empire in structures such as Santa Sophia in Istanbul, and then in Muslim architecture. Roman masonry architecture remains very heavy and the highest

Figure 4.3 York Minster.

levels of masonry development probably belong to the Gothic period in European architecture. At least in the view of the author of this chapter, the very pinnacle of the technical development of stone masonry was reached in the perpendicular style of architecture developed in England around 1350 (Figure 4.3). It is probable that modern tools and machinery could make some economies in masonry construction over the methods used in the fourteenth century, but it is doubtful if any real technical advances in design have occurred since that time. The structural forms developed by then were as near a perfect exploitation of a technology as it is possible to achieve.

Bricks

A development parallel to the development of stone masonry was that of fired brickwork. It has already been noted that fired brick was used in Mesopotamia to form a weatherproof skin to mud brick construction, but fired bricks could be used as a strong building material in their own right. The firing process makes them more expensive than mud brick but substantially stronger and more durable. Brick has never attained the architectural cachet of stone but its use became common-place in areas of good clay but little stone. Its structural properties are basically the same as stone masonry and so the same basic structural forms are used. It is a dom-inant material in domestic housing within the UK and many parts of Europe.

Timber

The disadvantage of masonry (stone or brick) is in the expense of using it to span gaps. We have seen that this can be done by arches, vaults and domes, but these are very expensive forms of construction and are only really viable in structures such as bridges or prestige buildings. The problem with arches was not only the direct expense resulting in part from the need to build a supporting structure for the arch or vault during construction, but also the construction depth required from the springing of the arch to the top, and the problems with resisting the out-ward forces developed at the supports of arches (see Chapter 3). A material that

Figure 4.4 Timber bridge.

Figure 4.5 Fifteenth-century timber-framed building.

could span by bending was necessary for more normal construction and, up until the Industrial Revolution, the only material that met this requirement was timber. Timber is the first material we have discussed that has good tensile strength as well as compressive strength, enabling it to resist bending. Also, its nature as the trunk of a tree meant that it was available in long, relatively thin sections that were ideal for beams. Timber was thus the ideal material for spanning medium-sized gaps. It was, and still is, the preferred material for supporting flat roofs and floors in small to medium-sized buildings. It could also be used for small span bridges (Figure 4.4).

For larger structures, timber could not be obtained in the necessary sizes or strengths, but here another property of timber could be exploited: it can be cut and jointed relatively easily, leading to the development of more complex structures formed from interconnected smaller members. The result was the development of the use of timber for framed structures and trusses. These types of structure were very important for the development of structures using more modern materials, but have been used in timber for centuries. England and other parts of Europe are particularly rich in timber-framed buildings developed from medieval times until relatively recently. They are particularly characteristic of areas that were forested and were short of other materials such as stone. Figure 4.5 shows an example of a fifteenth-century timber-framed structure. The areas between the framing were infilled with cheaper materials such as wattle and daub (mud or cow dung plastered onto a panel made of woven twigs). No doubt it was discovered at a very early stage that the tendency of the rafters in a pitched roof to spread could be stopped by providing a tie between them. From this insight it was a short step to the development of the trussed roof. Trussed roofs were developed fairly early in the Medieval period and evolved over the years into highly complex forms. Even today, it is not necessarily clear how these timber roof structures actually function. The triangulated truss, as we understand it, was probably a development of the Renaissance; in fact Palladio is often credited with the first clear illustrations of trusses. The principle of the truss has been discussed in Chapter 3 and it will be seen that some members work in tension while others work in compression. Timber is thus an ideal material for a truss, though the development of joint details, which could handle either tension or compression or both, could lead to some complexity. Timber trusses are still used very extensively and are the almost universal form of supporting structure for the roofs of domestic housing. Jointing has, however, become much more sophisticated in recent years.

Iron and steel

In 1779 Abraham Darby constructed his famous cast iron bridge at Ironbridge in Shropshire (Figure 4.6). This bridge was the public demonstration of a revolution that was to change the world. Abraham Darby had developed a means of using coal for the smelting of iron and the economic large-scale production of cast iron. This ready availability of relatively cheap iron is often considered as one of the prime factors leading to the Industrial Revolution. It is interesting that the first demonstration of this new ability was the construction of a bridge. The development of railways, roads and mills during the early years of the Industrial Revolution presented designers of structures with problems that were not readily solvable

Figure 4.6 Ironbridge.

Figure 4.7 A steel skeleton under construction.

using the traditional materials. Cast iron, wrought iron and, later, steel provided the solutions. Iron and steel have the same basic properties as timber in that they are strong in both tension and compression. They are both much stronger and much stiffer; however, due to the industrial nature of their production, they are also much more expensive than timber. Iron and steel are also heavier than timber. Even more than timber therefore, iron and steel were, from the start, used in frames and trusses where the material could be used in the most economical way.

In the early years of the development of railways, a substantial number of cast iron bridges were built. However, cast iron was largely abandoned after the collapse of the Dee Bridge in 1847. The problem with cast iron compared with wrought iron or steel is its brittleness. Brittle failures are something that engineers try to avoid, as they occur without warning and also provide no opportunity for forces and moments to redistribute to other stronger parts of a structure when failure of one member is imminent.

Wrought iron, which was a fairly expensive material, dominated the field until close to the end of the nineteenth century. The first major structure built using carbon steel was the St Louis Bridge over the Mississippi, completed in 1874. The first major steel structure in the UK was the Forth Bridge, completed in 1889. After this, the cheapness and convenience of rolled steel sections led to the fairly rapid displacement of wrought iron. The hot rolling process resulted inevitably in the standardisation of section sizes and shapes. Furthermore, the necessity for specialised equipment to cut, shape and drill rolled steel sections led to the members being formed ready for erection in a factory and then delivered to site ready for erection (Figure 4.7). This resulted in the potential for steel frames to be erected very rapidly. There has also been a tendency for the detailed design work to be carried out by steel fabricators rather than by consulting engineers.

A further fundamental development was the production of sheet and plate steel. Thin sheet metal was a major cladding material during the twentieth century, initially in the ubiquitous form of 'corrugated iron' and more recently in a variety of forms. Steel plate allows the fabrication of non-standard elements but, more critically, it can be used to form large box structures, which have been used with great success in recent years in bridge design.

In summary, in steel we have the first new structural material to arrive on the construction scene for, possibly, millennia. It is a highly versatile material of high strength and stiffness. The manufacturing process inevitably makes it expensive and, as a consequence, steel structures tend to be designed to minimise the quantity of material used, resulting in its use in frames and trusses or as thin sheet material stengthened either by the addition of stiffeners or by profiling the sheets. The manufacturing process and resulting properties also leads to the production of standardised sections, design by specialised fabricators, and the use of prefabricated elements.

Reinforced concrete

We shall now consider the second major new construction material to arrive over the last century or so. This is reinforced concrete.

Concrete is essentially artificial stone and, as such, has the same basic properties as stone. Its great advantage is that, as a man-made material, it can be poured into moulds of any shape where it sets, thus removing the necessity to form the material by carving, as is the case with stone. A further advantage is that its properties may be tailored to a considerable degree to meet different situations.

The basic ingredients of concrete are: gravel (usually stone in the sizes in the range of 5–20 mm), sand, water and cement. The cement is the only industrially produced ingredient and is used in relatively small quantities compared with the sand and gravel (typically about 15% by weight of the concrete). This makes concrete a very cheap construction material. The two basic types of cement are: *hydraulic cements* and *pozzolans*. Pozzolans were the earlier forms of cement and they can be found naturally as volcanic earths. If mixed with lime (calcium hydroxide) and water, pozzolans set to form a very effective concrete. Pozzolanic concrete was used extensively by the Romans: many of their great monuments were built by constructing a masonry skin and then filling this with concrete (the Colosseum, for example, is largely made this way). The most impressive Roman concrete building is probably the Pantheon in Rome. This is covered by a concrete dome 143 feet (43 metres) in diameter. This appears to have been cast in much the same way as we would today, by making a mould (formwork) and then pouring the concrete and, after hardening, removing the formwork to reveal the concrete surface. The long life of Roman structures illustrates the inherent durability of concrete. The second type of cement, hydraulic cement, reacts when water is added and requires no lime. The best-known hydraulic cement is *Ordinary Portland Cement* invented by Joseph Aspdin in 1811. This is now the most used commodity on Earth after water. Because hydraulic cements set rather faster than pozzolanic cements, they have largely displaced them; however pozzolans are used as replacements for some Ordinary Portland Cement in mixes for some uses.

Though concrete alone has great potential as a construction material, it shares one major weakness with stone. Stone is strong (often very strong) in compression, but, in tension, it is weak and brittle. If you have ever considered why there are so many columns in the Egyptian temple at Abu Simbel or why the columns in the Parthenon are so closely spaced, the reason is that it is impossible to make long span, reliable, stone beams. As you will remember from Chapter 3, high tensile stresses are developed within beams and this means that stone beams can only be short. Concrete has the same problem and cannot be used economically in any situation that requires it to resist bending. Wilkinson in England and Lambotte in France independently and at about the same time (in the 1850s) discovered how to circumvent this weakness. Wilkinson's 1854 patent for reinforced concrete explains how the steel ropes or bars were to be arranged in the formwork so that they finished up in the parts of the concrete members that would be subjected to tension under load. The concrete was thus used to support the compressive stresses and the steel to carry the tension. Steel bars are probably the cheapest means of supporting tension, while concrete is certainly the cheapest means of withstanding compressive forces. Reinforced concrete is thus an example of a composite material where ideal use is made of the materials.

Despite its invention in the 1850s, reinforced concrete was not really used to any great extent before the early years of the twentieth century. There was significant use in the years 1920 to 1939 but it was the Second World War that really led to the development of reinforced concrete as the pre-eminent structural material. This was mainly due to an extreme shortage of structural steel, which probably lasted from the war until the late 1960s. The result was that the great rebuilding throughout the world after the war was mainly done with reinforced concrete. It was probably only in the 1980s and 1990s that structural steel, due to a worldwide overproduction and a consequent major drop in price, started seriously to regain ground in the UK. In many, if not most, other countries, reinforced concrete still rules supreme.

Reinforced concrete turned out to be a greatly versatile material, able to be handled reasonably competently by a largely untrained workforce throughout the world. It is not, however, without its disadvantages. Two may be particularly mentioned. The first is its appearance. Concrete is a uniformly grey material, susceptible to staining from the environment, and large masses of exposed concrete can look deeply unattractive. The move to use exposed concrete in the 1960s led to some truly awful buildings that have given concrete a bad name that it has yet to live down. Concrete in buildings is nowadays usually covered discreetly by cladding. Concrete can, in fact, look stunning if designed, detailed and built correctly, but this requires inspired architecture and very careful construction. The second problem is with durability. As has been seen, concrete itself is highly durable and can last for centuries without serious degradation. There are some conditions that can lead to the degradation of concrete, and these will be discussed later, but they are relatively uncommon.

The real problem arises when steel is incorporated within the concrete since there are circumstances when this steel can corrode. Rust actually occupies a greater volume than the steel from which it is formed and, as a result, if the reinforcement corrodes, it tends to force off the surrounding concrete, leading to disintegration of the surface parts of the structure. There is also obviously a safety problem. Corrosion can be avoided by careful design and detailing but, in the days when reinforced concrete construction was booming, the understanding of the corrosion processes and the necessity to design to avoid problems were not fully realised. Consequently, much money has been spent in recent years on the repair of corrosion-damaged reinforced concrete structures.

Like structural steel, reinforced concrete is a highly versatile material; it probably comes closest of any major construction material to being a material that can be used for any form of structure. This breaking of the linkage between structural form and material properties is a major feature of reinforced concrete that designers may exploit in the development of economical or imaginative structures; more than with any other material, the possibilities of reinforced concrete are limited only by the designer's imagination.

Prestressed concrete

There is a second method of overcoming the weakness in tension of concrete. The principle may be seen by considering the problem of trying to lift a row of books (Figure 4.8). If we just lift the end books in the row, we shall lift only those two books. If, however, we provide a compressive force and 'squeeze' the line of books as we try to lift them, then we shall find that the books can be lifted. By providing an axial compressive stress we have converted our line of books into a book beam that can carry bending moment. The possibility of improving the performance of concrete by providing a longitudinal stress was recognised in the 1880s, but no practical working system of providing the longitudinal force developed. The reason was that concrete creeps under load. *Creep* is an increase in strain with

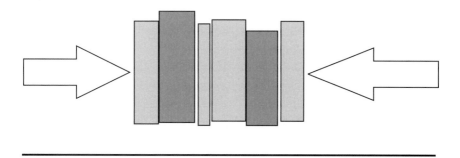

Figure 4.8 Lifting a row of books by pressing the ends (the principle of prestressing).

time in a material subjected to constant stress. The effect of the creep is to reduce the magnitude of the longitudinal force with time, resulting in failure of the beam. In the end, this problem was solved by Freyssinet in France in the 1920s after many years of experiment. He realised that the creep problem could be over-come by using high-strength concrete, with steel wires of very high strength to provide the force. Two basic systems of *prestressing* (as this system of imposing a longitudinal compression on concrete members came to be called) developed: pre-tensioning and post-tensioning.

Pre-tensioning is usually a factory process because it requires the facility to stretch wires and hold them under tension for some time. High tensile wires are stretched along the length of the casting floor. The formwork for the members (the mould) is constructed around the wires and then filled with concrete (Figure 4.9a). When the concrete has set and gained sufficient strength, the wires are cut. Since the wires should have become bonded to the concrete, this transfers the tension to the concrete. The resulting prestressed beam is then transported to site and erected.

Post-tensioning is usually a site process and is used for larger or more compli-cated structures. The structure is cast with ducts (tubes) set in where the prestressing wires are required. When the concrete has hardened, high tensile steel wires or cables are threaded through the ducts and anchored at one end. A jack is fixed to the wires at the other end and the wires are tensioned (Figure 4.9b). Once this has been done, anchors are fixed on so that when the jack is removed the tension remains in the wires. The ducts may now be filled with grout (liquid mortar) to protect the wires and bond them to the beam, or may be left ungrouted so that, if necessary, the wires can be removed for inspection or replacement in the future.

The pre-tensioning process lends itself to the production of numbers of similar units under factory conditions. The nature of the process tends to favour straight members. The elements need to be small enough to transport from the factory to

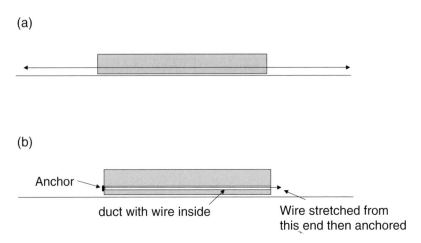

Figure 4.9 Methods of prestressing: (a) pre-tensioning; (b) post-tensioning.

the site. Typical pre-tensioned products are: railway sleepers, standard beams for bridges, and floor planks for making precast floors for commercial and residential buildings.

Post-tensioning lends itself to use in structurally much more exciting situations. It is probably most commonly used in large bridges. One way it is often used is in the construction of segmental bridges. This type of bridge is made up of units that are precast, usually on or near the site because of their size. Each new segment is hoisted into position against a previous segment and then prestressing tendons are threaded through the unit and connected on to the previous unit. The tendons are then tensioned to pull the unit tightly against the previous units. This is just the same procedure as lifting a pile of books by providing a compressive force on the ends.

Discussion

The objective of this chapter has been to show how different materials tend to be suited to different forms of construction. It is certainly not the intention to suggest that one material is necessarily better than another. Mud or earth, for example are not inferior materials; for building a dam, earth is probably the most commonly-used material. Some materials quite severely limit the forms of structure for which they can be used. Masonry, for example, is very weak in tension or bending, but remains one of the commonest structural materials and one which, in the past, has been used for some of the most beautiful buildings produced by humankind.

What can be said is that the two new materials, structural steel and reinforced or prestressed concrete, are probably the most versatile and impose fewest limits on what may be done with them structurally. Nevertheless, there are structures for which concrete is the obvious choice, and structures for which steel is the obvious choice. It is hardly possible, for example, to imagine any material but concrete being used for an arch dam. The obvious material for a particular structure may not remain the same over time. For example, 25 years ago, almost all portal frame sheds would have been made from concrete. Nowadays steel is universally used. This change has little to do with the structural possibilities of the two materials, but is the result of changes in the relative price of the materials. In 25 years' time the obvious choice could have changed back to concrete.

The major change for designers today compared with the past is that almost any structural material is available anywhere in the world. A designer is therefore not forced to use the materials available locally but is free to choose the most appropriate from the whole catalogue of possibilities. This might suggest that modern structures should be greatly superior to those built in the past. The reader may like to consider whether this is so.

Further reading

IABSE (1981), *The Selection of Structural Form* , Proceedings of IABSE Symposium, IABSE, London. Many papers are worth reading, including: Mainstone, R. 'History of climatic influences in building design', Fitch, J. M. 'History of climatic influences in building design' and Happold, E. 'Materials and components', IABSE, London.

Chapter 5

Loads and loading

Introduction

One of my friends – an eminent engineer – once remarked that loading was one of the biggest unknowns in engineering. This rather facetious observation is more profound than it may first appear. For instance when we use a uniformly distributed load in design, it should be appreciated that the loading may be far from uniform and its magnitude may not always be what is encountered in practice.

For example, when a room is packed with people to such an extent that shuffling ceases, the area occupied per person is 0.26 m^2 and it yields a loading of approximately 2.4 kN/m^2, which is the design-superimposed loading for offices! On the other hand, we know that some parts of an office – usually small areas – are heavily loaded with paper, files or equipment. Figure 5.1 illustrates pictorially some of the loading encountered in practice.

The point of this illustration is to show that arriving at a satisfactory design loading is not straightforward. Values used in practice have been known to give satisfactory results and are accepted on the basis of experience. The aim of the discussion in this chapter is not to provide definitive values for different types of loads that can be found in Codes of Practice, but rather to give some insight into the nature of loading, which would lead to an intelligent approach to design.

Sources of loads and classification

Gravity loads

Structures are generally required to support items which have *mass*. The gravitational force of these masses (the *weight*) acts on the structure. Self-weight of the structure and finishes are called *permanent actions*. (The term 'actions' is used in Eurocodes and is an appropriate general description to cover all types of loading.) Loading caused by the occupancy of the building is the other major source of loading, commonly called *superimposed loading* or live loads or variable action. In general this fluctuates with time. Although partitions in buildings may look

Figure 5.1 What it takes to impose a loading of 2.5 kN/m^2 (above); typical office loading of about 0.7 kN/m^2 (below).

'permanent' when built, they could be modified, and their weight should strictly be classified as a live load.

Climatic loads

Another major source of load is the climate. Wind loading and snow loading are prime examples. These two loads are discussed further later in the chapter. At this stage it suffices to say that wind can act in any direction, and the loading on a structure can be normal or tangential to the surfaces, producing horizontal and vertical loading.

Roofs will need to be designed for the weight of snow likely to be deposited in the winter and the possible accumulation caused by drifting.

Icing is another climate-induced load that needs consideration (in the British Isles and similar climates) in the design of some types of structures such as cable-supported structures and towers. A television mast in Elmley Moor, Yorkshire collapsed owing to icing on the cables. A number of questions arise: the weight of ice that can form, its shape, wind speeds that are likely to occur during and after ice deposition and the frequency of this phenomenon.

Another form of climate-induced load arises when movements caused by temperature changes are restrained.

Another type of 'load' induced by climate is rain. The design of storm drainage requires knowledge of the intensity of rainfall (mm/hr). Statistics on the duration of storms and the frequency of storms of particular intensities will assist in sizing sewers and assessing the risk of flooding.

Hydrostatic loads and soil pressures

Hydrostatic loads occur in many forms. In the design of water tanks or dams, the lateral loading imposed by water is the major design load. In the design of tanks located above ground, the weight of the water also needs supporting. Ground water causes lateral loads on basement walls. It also causes buoyancy of the structure. The main uncertainty with ground water is the level of the water table, which can be ascertained only by installing piezometers (stand pipes) and monitoring over a period.

Soils cannot generally stand up vertically by themselves. The surface of a stable heap of soil will be at an angle $< 90°$ to the horizontal. This is referred to as the *angle of repose* and depends on the characteristics of the soil, particularly the angle of internal friction, cohesion and moisture content. When forced to a vertical profile i.e. when retained, soils, just like water, exert lateral loads on the retaining structures. The magnitude of the load depends on the type of soil – whether it is cohesive (clay) or granular (gravel/sand), geotechnical properties mentioned earlier, the density of soil, and the height of retention. It also critically depends on whether the retained structure is allowed to move or 'give'. When the wall gives the pressure drops to what is called its 'active' value from the pressure at rest. Similarly, when the structure pushes against the soil, the soil exerts 'passive' pressure, which is considerably greater than the active values. Passive pressure is 'available' to resist lateral loads and can be mobilised only when a slight movement of the structure takes place.

Lateral loading on the walls of containers for granular solids (referred to as silos or bins) is still not perfectly understood. While the principles of soil mechanics can be used when the contained material is at rest, large transient forces or dynamic effects occur during the discharge of materials from a silo or bunker. Horizontal loading can be increased by a factor of 3, and many failures have occurred in practice where this special factor had not been addressed.

Accidental loads

Accidental actions arise out of unintended events such as explosions, impact, fire or consequences of human error. It is axiomatic that structures are designed to resist all foreseeable loads during their construction and use. Strategies to limit the damage caused by accidents include avoidance or elimination of the hazards, selection of a 'robust' structural form with low sensitivity to the hazards considered, and providing the structure with adequate ties and ductility to enable survival of the structure. Providing alternative load paths, in the event of a main support being lost in an accident, is also a legitimate measure.

Indirect actions

There are some classes of actions caused by movement. These are called *indirect actions*. The simplest example is the force induced in a structure when expansion or contraction caused by temperature fluctuations, is resisted. Note that an unrestrained free movement does not induce any forces. Take the case of a steel beam of cross-sectional area A cm^2 and length L m. Let us assume a temperature increase of T degrees Kelvin. The free expansion of the beam will be (αLT) where α is the coefficient of thermal expansion (Figure 5.2).

The *strain* is the ratio of the expansion to the original length, i.e. αT. If the beam is prevented by rigid buttresses from changing its length, then the expanded beam has to be pushed back by αLT to its original length L, thus causing a compression in the beam. The compressive stress can be obtained from the relationship: stress $\sigma \div$ strain ε = modulus of elasticity E. Therefore the stress $\sigma = E\varepsilon = E\alpha T$ and the force in the member is (stress × area) = $\sigma A = E\alpha TA$.

Note that the force is independent of length L and, for a given material, is proportional to the temperature change and the area of the member. The magnitude of the movement is directly proportional to length.

As an example, for $T = 30$ K and $A = 29.8$ cm^2 (152 × 152 × 23 kg Universal Column): stress = $(2.1 \times 10^5) \times (12 \times 10^{-6}) \times 30 = 75.6$ N/mm^2.

The member should be checked to ensure that it can carry this stress, taking into account the 'slenderness' of the member where its length will be a consideration. The buttresses should be able to withstand the induced force of $\sigma A = 225$ kN.

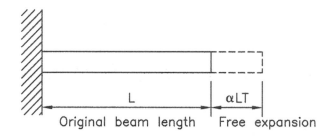

Free movement ⟹ No internal forces

(a)

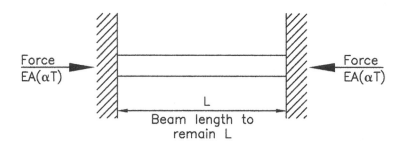

Movement constrained ⟹ Internal force

(b)

Figure 5.2 Movement and forces: (a) free movement: (b) constrained movement.

In practice unyielding buttresses are rare, and any 'give' will result in a reduction in the axial stress. This is because the beam now has to be pushed back by $(\alpha LT - \Delta)$, where Δ is the movement of the buttress. The strain, and hence the stress, will thus be reduced.

This rudimentary illustration is applicable to all structures and all materials. It also shows that allowing movements is far better than resisting them. It requires the structure to be broken up with 'movement joints' at suitable spacing to control the amount of movement. This is why long-span structures such as bridges incorporate 'bearings' that permit sliding. If a massive bridge structure is prevented from moving, the forces generated will be colossal (remember that the force is proportional to the cross-sectional area of the member), and bridge piers will need to be designed for this huge lateral force. At a smaller scale, brick cladding usually incorporates joints at 12 m centres to avoid building up unnecessary in-plane

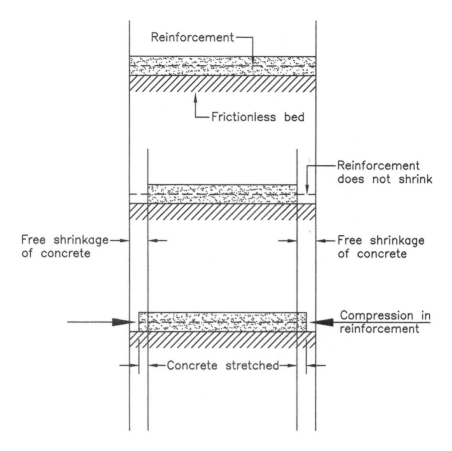

Figure 5.3 Forces due to shrinkage.

forces which in turn may have undesirable consequences such as bowing and spalling.

The cause of movement need not only be temperature. Concrete structures 'shrink' after they are cast as some of the moisture in the concrete dries out. It is in fact referred to as *drying shrinkage*. Again, if the movement is unrestrained there will be no stress or force to be resisted (Figure 5.3).

Restraint can arise from a number of sources, such as friction between ground slabs and ground, and the presence of columns and walls. It is interesting to note that the presence of reinforcement itself is a restraint, as the reinforcement does not shrink! Thus even where a reinforced slab is cast on a frictionless bed, shrinkage will cause tension in the concrete. The bond between the concrete and reinforcement will force them to move the same amount. The concrete has to be 'stretched' from its freely-shrunk location and at the same time the reinforcement

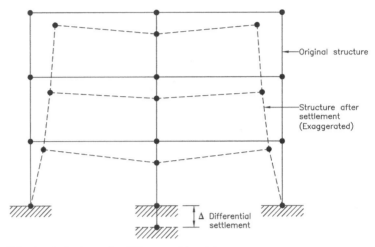

Fully pinned structure ⇒ Little or no internal force

(a)

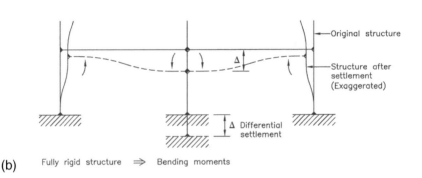

(b)   Fully rigid structure ⇒ Bending moments

Figure 5.4 Forces caused by settlement.

has to be compressed. If no reinforcement were present the mass concrete would simply have shrunk with no internal forces induced.

Another type of indirect action that gives rise to internal forces in a structure is settlement or heave of foundations. Consider a two-bay multi-storey frame, such as in Figure 5.4(a). Let us assume a differential settlement of Δ between the foundations under the central column and the outer columns. If the joints between the columns and beams at various floor levels were perfectly 'pinned' then the result will be a slightly altered configuration of the structure, with the members maintaining their original length, and the floor beams sloping towards the centre. However, if there are rigid connections between the columns and beams, then this structural continuity will give rise to a 'flexing' of the beams (Figure 5.4(b)). This will result in equal and opposite bending moments at the ends of the beams. It can be shown that the magnitude of bending moment is $6EI\Delta/L^2$. This shows that for a given size of member (property EI) the bending moment is proportional to the magnitude of the settlement and inversely related to the square of its span length. Thus long members can flex without causing significant internal forces but short, stiff members will generate considerable forces.

A similar problem will arise under fire conditions in rigid frames with significantly different column sizes. As a general rule, the temperature rise is proportional to the mass of the member. If adjacent columns thus undergo differential thermal expansions, bending moments in the beams as described above will result (of course the sign of the bending moments will be opposite to those caused by settlement).

Dynamic loads

Dynamic actions can cause significant acceleration of the structure. There are many examples of dynamic actions: harmonic loads imposed by machinery, pulse loads caused by impact, transient loads such as explosions, repeated transient loads such as pile drilling, transient random loads (earthquakes), wind and waves, bell ringing, dancing, etc. The response of the structure will depend on the properties of the structure as well as the characteristics of the load. In this context the main properties are the natural frequency and the damping of the structure. As a general rule, stiff structures possess high natural frequency. When the structure is sufficiently stiff, dynamic actions may be treated as static actions termed as a *quasi-static load*. In some structures that are 'mildly' dynamic (i.e. those that fall in the grey area between static and dynamic), the loads could be augmented or magnified and treated as a static action. In dynamic analysis, equations of motion of the structure (taking into account the inertial forces and applied loads) will need to be solved. Boundary conditions will determine the constants in the solution. The procedure is generally complex.

Fatigue loading

This is essentially a materials problem. Where structures are subject to pulsating or alternating loads, their cumulative effect leads to a failure called *fatigue failure*. You must be familiar with the failure of paper clips when bent forwards and back a few times. It is usually a local material failure initiated at a stress raiser such as a sharp notch or a crack. For a given mean stress level it has been found that the number of cycles N at which failure occurs decreases as the stress range S increases (Figure 5.5).

The failure values are determined experimentally, and a family of $S - N$ curves is set up for various mean stress levels. The likely number of cycles of fluctuation of stress is considered in conjunction with the stress range at a given mean stress level. Fatigue is particularly critical for welding details when stress fluctuations occur e.g. bridge structures, gantry girders or other lifting gear.

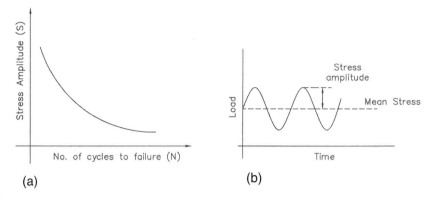

Figure 5.5 Fatigue strength.

Fire load

This is different in character from the other loads discussed above. It represents the amount of combustible materials. Fire load is usually expressed in MJ/m^2 of floor area and clearly depends on the type of occupancy or use of the building. Fire load is not used to calculate the mechanical behaviour but it is one of the factors that determines the building's *fire severity* (the maximum temperature of the structure in fire conditions), which in turn has an effect on the strength of structural members. All materials lose strength at very high temperatures and therefore generally need fire protection. The type and thickness of fire protection can be calculated if the fire severity is known.

Statistical concepts used in conjunction with loading

Introduction

Modern designs are carried out using *limit state* concepts (see Chapter 7). In this approach, the designer verifies that, in all relevant design situations, the limit states are not violated when the appropriate values of actions (loads), material properties and geotechnical data are used in the design models. Limit states are conditions beyond which the structure no longer satisfies the design performance requirements. Many limit states could be considered, but they all broadly fall into two categories: *ultimate limit states* (ULS) and *serviceability limit states* (SLS). The former are involved with collapse or similar failures and the latter are associated with conditions such as deformation, cracking and vibration. Clearly, different loadings must be used for each verification. The ultimate limit state concerns the stability and safety of the structure under the most extreme loading that is likely to occur. In serviceability verification the loading involved will be considerably less, reflecting the service conditions.

Quantitative data of loads will be required to translate the above concepts into practical calculations. The process of determining the values of loads is, in practice, a combination of measurements and calculation (self-weights), observation and measurements (wind and snow), experience and surveys (floor loadings) and judgement (some accidental actions). Usually calculation, measurements and observation will yield data, which could be analysed statistically, and the results presented in statistical terms. However, Codes of Practice usually treat values obtained by other methods (such as nominal values prescribed by clients) in the same way. Some background discussion on the most commonly used approaches is presented below.

Characteristic values of action

Where fixed on a statistical basis, this corresponds to a value that has a prescribed probability of *exceedence* in a reference period. For permanent actions (self-weight etc), where the variability is relatively small (coefficient of variation less than 0.1 according to Eurocode 1), the mean value is taken as the characteristic value G_k (suffix 'k' is commonly used to indicate characteristic value). For variable loads, we are interested in the magnitude of load that will be exceeded with a known probability; this probability of exceedence (p) is usually 0.05 in a year, although in the UK wind and snow loads are prescribed with a lower probability of exceedence of 0.02 in a year.

Sometimes the term *return period* or *recurrence interval* (R) is used in this context. It represents the mean duration between occurrences of values in excess of the characteristic values. It must be remembered that no periodicity is implied, i.e. if the characteristic value is exceeded this year, there is no guarantee it will not occur next year. It represents an average period in an indefinite trial when $R = p^{-1}$. As it can give rise to misunderstanding, the use of the term 'return period' is discouraged and 'probability of exceedence in one year' is used instead. The normal convention for the characteristic value of a variable action is Q_k.

In design, representative values of variable action other than characteristic values are often required. These will now be discussed.

Combination value

The *combination value* $\Psi_o Q_k$ is chosen so that the probability that the effects caused by the combination will be exceeded is approximately the same as that for the characteristic value of an individual action.

Take the case of a structure that is acted on simultaneously by two loads Q_1 and Q_2 with their own probability distributions. How do you combine these? Do we add the respective characteristic values ($Q_{k1} + Q_{k2}$)? That will be the answer if the two loads are correlated, i.e. occurrence of one load leads to the occurrence of the other. In practice there are many uncorrelated loads. In these cases the aim is to produce a combination, the joint probability of occurrence of which is the same as that of a single action. The simplest method (used in Eurocodes) is to consider one action at its characteristic level and the other at its combination value. Thus $(Q_{k1}+\Psi oQ_{k2})$ and $(\Psi oQ_{k1}+ Q_{k2})$ are possible values. More than two loads are treated similarly by taking each load in turn as the dominant load.

Frequent value

The *frequent value* $\Psi_1 Q_k$ is determined so that either the total time within the reference period during which it is exceeded is only a small part of the reference period, or the frequency of its exceedence is limited to a given value.

When the distribution of the load is known over a reference period t, the frequent value is chosen such that it is exceeded over only a small fraction of that period. The Eurocode value for building structures is 0.01. Multiplier Ψ_1 (< 1) is used to arrive at the value of frequent value of action. It may be assumed that the frequent value represents the likely level of loading that might be encountered in a structure in use.

When considering the response of a structure at ultimate limit state to accidental loads, frequent values of other variable actions rather than their full characteristic values are included, as accidents occur in a building in use. Most serviceability checks are carried out using the frequent values of variable actions.

Quasi-permanent value

The total time over which *quasi-permanent values* $\Psi_2 Q_k$ are exceeded will be in a greater proportion than for frequent values. In Eurocodes the fraction of the time when quasi-permanent values is exceeded is taken as 50%.

Many types of variable actions have a permanent component, e.g. floor loadings in offices or domestic premises. On the other hand, wind and snow loads have none, at least in the UK climate. Long-term settlements of foundations and long-term deflection of timber or concrete floors are calculated using permanent loads (dead loads) and quasi-permanent loads.

The above concepts are pictorially presented in Figure 5.6.

Discussion of some types of loads

Floor loads

In the UK loads are specified in the British Standard BS 6399:Pt 1. Loads in BS Codes have remained relatively unchanged over many years. In the 1970s Mitchell and Woodgate of the Building Research Establishment carried out surveys of loads in domestic buildings, office buildings and retail premises (BRSCP 3/71, 2/77, 25/77). The weights of the items on the floor were measured. These observations were then used to calculate the average load over notional bays of different sizes. Frequency distributions of load were then calculated for each bay size.

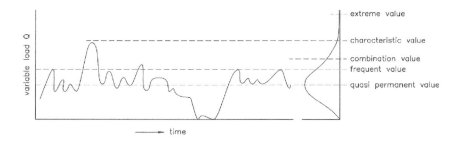

Figure 5.6 Statistical description of loads.

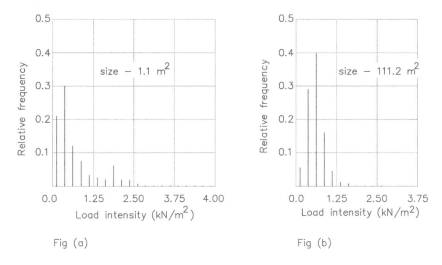

Fig (a)   Fig (b)

Figure 5.7 Relative frequency of load intensity in offices.

As an example Figure 5.7 shows the distribution for bay areas of 1.1 m² and 111.2 m² for office buildings. Mitchell and Woodgate also calculated the 99.9% probability loadings for different bay sizes in offices, making allowances for twelve changes of occupiers. The results are shown in Table 5.1.

The surveys show that the intensity of loading decreases as the area of the bay increases. Codes of Practice allow for this by reduction factors for loads on a single member (say a beam) from a large tributary area. There is another reduction factor applied to imposed loads supported by vertical elements (columns and walls) in multi-storey structures. This is to allow for the reduced probability of all the floors being loaded at characteristic loads simultaneously. The reduction does not apply to storage loads or loads specifically derived for the particular use or loads from plant and equipment. All these loads can be present all the time.

Table 5.1 Load intensity levels for office floors other than lowest basements and ground
floors after twelve changes of occupation

Size group	A	B	C	D	E	F	G	H	J
Mean area m^2	1.1	1.4	2.4	5.2	14.0	31.2	58.0	111.2	192.2
Probable load intensity kN/m^2									
Mean		3.55	3.24	2.81	2.36	1.73	1.43	1.26	1.16	0.99
99.9% probability		17.38	10.87	8.04	5.31	4.31	3.54	3.45	3.18	2.34

Roof loads

In the UK, loads on roofs mainly arise from snow, access for maintenance, the roof acting as a fire escape route, and wind. The standard BS 6399:Part 3 deals with the first and it is accepted that the minimum snow load allowance in the code will be sufficient to cater for loads arising from occasional access for routine maintenance. When the roof acts as a fire escape route, it should be considered as a floor. Wind loading is discussed later.

The rest of this section is concerned with snow loading. The deposition and redistribution of snow on roofs are complex phenomena. Fresh snowfall occurs in a range of densities depending on meteorological conditions. Some snow is very dry and powdery, whereas other snowfalls may produce very wet, sticky (and therefore dense) snow. Once snow has fallen, subsequent winds and temperature variations, together with consolidation with time due to self-weight, cause changes to the initial fresh snow density. The loadings given in the code have been derived statistically from snow depth records maintained by the Meteorological Office, and conversion factors (density relationships) are also derived statistically. A map of basic snow load on the ground is presented in the code. Variation of snow load with altitude is recognised.

Snow is naturally deposited in many different patterns on a roof, depending upon the wind speed, the wind direction, the type of snow, the roof geometry and characteristics of any obstructions (Figure 5.8). Therefore two primary loading cases need consideration:

- a uniformly distributed layer of snow with no wind
- those resulting from redistributed snow likely to occur under windy conditions.

The latter has been assumed to be governed by three factors:

- The space available in which the drift could form
- The amount of snow available on the roof for redistribution into the drift
- The maximum size of drift considered appropriate to the UK.

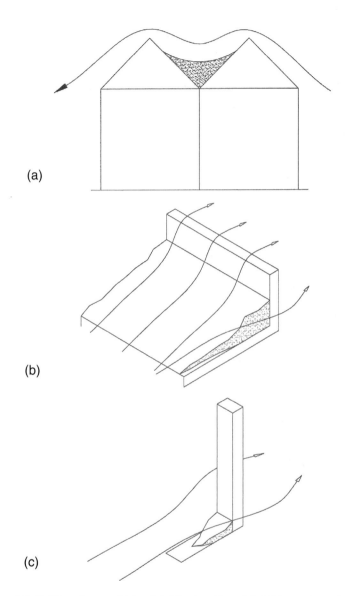

Figure 5.8 Drifting of snow: (a) local drifting in valleys; (b) drifting against a long obstruction; (c) drifting against a local obstruction.

For various roof shapes the code provides coefficients for snow load shapes corresponding to the three limiting conditions noted above. As each is the maximum value, the least of the three will be the critical design condition. Another point, which should be borne in mind in this context, is that the UK is governed by

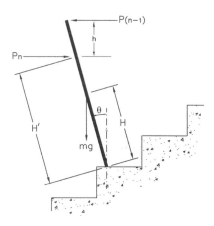

Figure 5.9 Forces acting on a handrail by a person leaning on it.

'single snow' events rather than 'multiple snow' events. In the UK it is assumed that snow from one weather system is redistributed into local drifts by wind from one direction. The shape coefficients will be different in other climates.

Horizontal loads on handrails, barriers, etc

Consider a person leaning against a handrail, as shown in Figure 5.9. By taking moments about the foot, the horizontal reaction on the rail can be seen to be

$$P_n = mg\left(\frac{H}{H'}\right)\tan\theta.$$

If the person is pushed from behind, as shown in the figure by a force $P_{(n-1)}$ then

$$P_n = mg\left(\frac{H}{H'}\right)\tan\theta + P_{(n-1)}\left[1+\frac{h}{H'+\cos\theta}\right] \qquad .$$

The force $P_{(n-1)}$ can be the result of several spectators leaning at an angle θ and pushing against the person in front. This simple model illustrates the principles of force generation in safety barriers and rails. Considerable research has been carried out into the likely forces on safety barriers in the context of crowd safety in spectator terraces, where deaths have occurred due to crushing. Clearly the crowd density, the terrace inclination and the spacing of barriers will all influence the magnitude of the force. Major spectator facilities will require certificates by the appropriate authority and they will specify the loadings to be used. BS 6399:Pt 1 contains provisions for horizontal loads for parapets, barriers, balustrades, etc.

The severity of loading depends on the type of use of the building and the consequences of failure.

Vehicular barriers are designed to absorb the kinetic energy of moving vehicles as strain energy in the barrier and the vehicle itself. Force on the barrier will thus be $0.5mv^2/(\delta_c+\delta_b)$, where m is the mass of the vehicle travelling at a speed of v and δ_c and δ_b are the deformation of the vehicle and the barrier respectively.

Wind loads

Characteristics of wind

All structures above ground are subject to the effects of wind. The main effect of concern to civil engineers will be the loading on buildings, bridges, towers and other structures. Windiness is also relevant to other problems in engineering such as environmental comfort, wind energy, air quality and dispersion of air pollution. This section is concerned only with mechanical loading.

Assessment of wind loading is undertaken in two stages. The first is the meteorological data used to predict the incident wind speed on structures. The second is concerned with the wind flow over and pressures applied to the surface of structures.

Wind is the source of energy that drives the boundary layer next to the Earth's surface. This layer is some 2.5 km high above the Earth's surface and is that part of the atmosphere affected by the surface characteristics of the Earth. The energy is dissipated by friction with the rough surface terrain, of which the structure is an 'element'. The energy flow is experienced by structures as imposed wind load.

Wind is a highly dynamic load. Wind velocities measured at any fixed point will vary randomly with time Figure 5.10(a). This feature is called turbulence, and its main source is the surface roughness of the Earth. The building on which the incident wind acts will also interact with the wind and modify the flow pattern and create additional small-scale turbulence itself. The wind speed trace shown in Figure 5.10(a) can be separated into two components. The wind velocity shown in Figure 5.10(b) is referred to as the mean speed and contains speeds up to a frequency of 6 cycles/hr (or 10 min/cycle). Figure 5.10(c) shows the fluctuating part, and contains the higher frequency information.

In general, for structural loadings the speeds averaged over short duration (e.g. 3 seconds) are relevant, as it is known that structural elements respond to gusts of such short duration. In other words it will be unsafe to base the structural design on mean wind speed averaged over a relatively long period and ignore the much higher speeds of short duration gusts (and hence the loads caused by them).

Although wind loading is highly dynamic, most structures are designed with it treated as a 'quasi-static' load. In this procedure it is assumed that all fluctuations of load are due to the gusts of the boundary layer and not due to other building-generated turbulence. The structural response to the gusts is assumed to be similar to the response to mean flow. This approach is valid provided the structure is

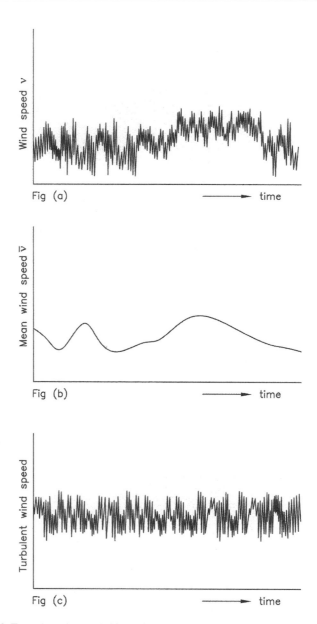

Figure 5.10 Typical wind record: (a) wind speed trace; (b) mean wind speed; (c) turbulent wind speed.

sufficiently stiff. The UK Code of Practice BS6399:Pt 2 contains a table at the very beginning of the document to test whether the quasi-static approach is applicable to the structure under consideration. A structure that falls outside the is too

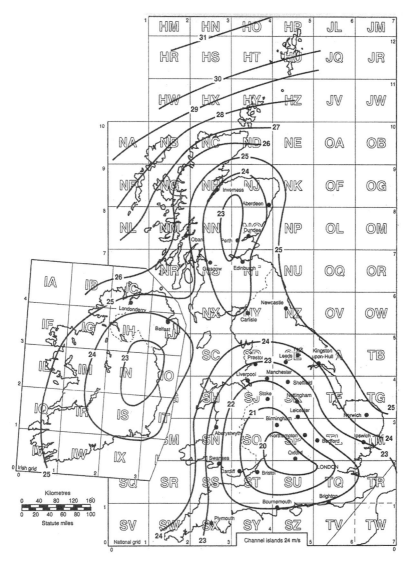

Figure 5.11 UK wind speed map (reproduced from BS 6399–2, 1997).

flexible and requires a full dynamic analysis, allowing not only the turbulence of the boundary layer but also for turbulence generated by the structure itself.

Wind speeds

Wind speed contours are normally presented in Codes of Practice (Figure 5.11) and designers use this basic speed as a starting point. In the UK multipliers for

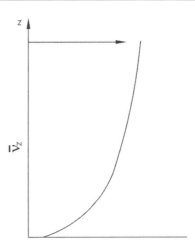

Figure 5.12 Variation of wind speed with height.

altitude, direction, season and risk probability correct hourly mean speeds taken from the map. When the site is located on a hill or escarpment, wind accelerates over the obstruction and additional wind forces could occur on a structure depending on the slope of the hill and the location of the site in relation to the crest. It has also been found that in the UK the strongest winds are from the south-west and the weakest are about 73% of this value and from the north-east. For structures that are exposed to wind during specific subannual periods it may be possible to reduce the design wind speed, knowing the exact month or months when the structure is to be put up. The quiet months (for wind loading in the UK) are May, June and July. The basic wind map contains speeds that have an annual risk of exceedence of 2%. If it is desired to vary this risk, a probability factor can be used (e.g. 1.26 for a risk of exceedence of 10^{-4} used for some nuclear installations) to modify the basic speed.

It will be instructive to look at the basis for deriving the basic wind speeds. In the UK the Meteorological Office holds the wind data derived from continuously recording anemographs normally exposed at a height of 10 m above ground in open, level terrain. There are about 130 stations. The aim is to obtain the basic wind speed, which has an estimated risk of exceedence of 0.02 in any one year. This implies a return period of 50 years, in the long run. How does one predict this with data that covers a period of less than 50 years? Predictions are carried out using statistical theories, particularly those concerned with extreme values.

Variation of wind speed

The variation of wind speed with height can be described using a simple power–law relationship (see Figure 5.12) such as

$$\overline{V}_z = \overline{V}_{10}\left(\frac{z}{10}\right)^\alpha$$

where $\overline{V}_z$ is the hourly mean speed at height z m, $\overline{V}_{10}$ is the mean speed at 10m above ground in open terrain and   is the power law index.

In any given terrain wind speed at 10 m height can be expressed as

$$\overline{V}_{10} = R\overline{V}_B$$

where $\overline{V}_B$ is the value applicable to open terrain. Codes of Practice provide values for α and R for different terrain descriptions, ranging from seac oasts to suburban regions in towns.

When the surface roughness is large and closely packed, as in towns, the wind flow tends to skip over the tops of the buildings. It is as if the wind sees the ground level as the average height of the buildings. This is called *zero plane displacement* and can be taken into account in calculating the effective height of buildings in towns.

It should be appreciated that the mean wind speed increases with height above ground and decreases with increasing ground roughness. The turbulent component, however, decreases with height and increases with ground roughness. The atmospheric boundary layer is in a constant state of adjustment as it passes over varying ground roughness. The upwind extent of each kind of ground roughness (termed *fetch*) has to be of the order of 100 kilometres for the wind to reach equilibrium profile. When the terrain changes, the change of wind profile is gradual. In BS 6399:Pt 2 the factors defining wind speed for different effective heights in town, assume a minimum two kilometre fetch of town terrain upwind. The above standard also presents an alternative method of obtaining site wind speed, taking into account the actual distance of the site from the sea and distance into town. This could be performed for different directions also allowing for direction factors. Alternatively, a conservative value can be estimated using the southwest wind and minimum distance from sea.

Wind flow over building surfaces

The above discussion provides the background to the procedures used in codes for arriving at design wind speed v. The basis for the calculation of loading q is $0.5\rho V^2 C_p$, where   is the density of air (1.23 kg/m^3) and C_p is the pressure coefficient. C_p is thus the ratio of the actual wind load experienced at a location to the kinetic energy of the incident wind.

The pressure coefficients are obtained from wind tunnel tests. Some general discussion of flow patterns around 'bluff bodies' will be useful to clear up some common misconceptions. Windward face, side faces, roof and the rear faces of a simple rectilinear building are considered.

A body is aerodynamically *bluff* when the streamlines do not follow the surface of the body, but detach from it, leaving regions of separated flow and a wide-trailing wake. [The opposite of a bluff body is a *streamlined* body, in which the flow lines remain tangential to the surface everywhere.] Take the case of a rectangular building with a flat roof. Because the wind profile in open country follows a power law variation with zero value at ground level, many engineers expect the pressure distribution also to vary from zero at the base of the wall to a maximum at the top. This will be so if the wind is not allowed to escape over the roof and around the sides. At about two-thirds of the building height there is a point of stagnation. Above this point the wind flows up the wall, and below it, down the wall. The flow when it hits the ground has more kinetic energy than the incident wind at this level (based on power law variation). It therefore moves forward against the wind and loses energy until it comes to rest on the ground at some distance in front of the wall. It now rolls up as a vortex. This is illustrated in Figure 5.13(a) and the resulting pressure distribution is shown in Figure 5.13(b) in which the pressure coefficients shown relate to the wind speed at the top of the wall.

As regards the side walls, the flow over the windward face separates from the building surface along the line of the corner with the sides. The vortex discussed above escapes around the side and is significantly faster than the incident wind at that level, thus making the flow on the sides more or less uniform. The flow is detached from the sides, implying negative pressures or suction with the maximum value at the leading edge.

Similarly, the separated flow over the roof surfaces tends to produce negative pressures (suctions) over the surface, which are pronounced at the windward edge and gradually reduce. If the building is long enough, the separated streamlines may re-attach themselves to the roof surface.

The *wake* is the term used to describe the flow behind the bluff body. Figure 5.14 shows a typical pattern.

It should also be obvious from the above that the pressures tend to be more or less uniform from top to bottom, particularly in zones of negative pressure. This is the reason why it is erroneous to divide the building into horizontal slices or parts and calculate wind speeds for each part (the so-called 'division by parts'). Limited application of division by parts is permitted on windward faces, but the height of the slice needs to be at least the cross-wind breadth of the structure.

Flow around rounded bodies is more complex. There is generally a sharp change in the flow pattern when the speed reaches a critical *Reynolds number* (a non-dimensional parameter used in fluid mechanics with a value of $6 \times 10^4 Vd$ in normal weather conditions, where V is the wind speed and d is the diameter of the cylinder). In the supercritical range (high values of Reynolds number) the wake is much narrower than for the subcritical range, and the drag coefficient (equivalent to pressure coefficient) drops from 1.2 to about 0.5.

In calculating the net load on a surface, note has to be taken of the external and internal pressures on the surface. The internal pressure in a clad building depends on the relative ease of inflow and outflow of air and on the pressure drop

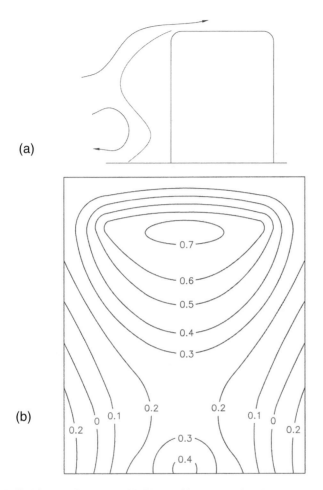

Figure 5.13 (a) Wind flow over a bluff body; (b) pressure distribution on a windward-facing wall.

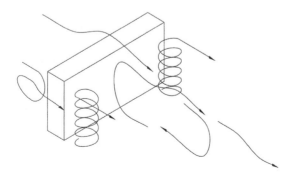

Figure 5.14 Wind circulation behind a slab block.

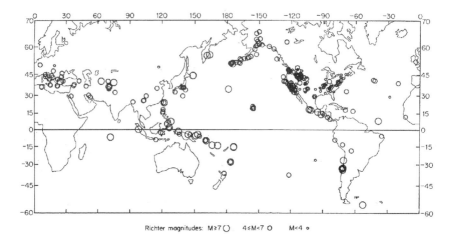

Figure 5.15 World seismicity: locations of earthquake epicentres (1983–86).

Source: Smith, J. W. (1988) *Applications in Civil Engineering*, Chapman Hall.

across any apertures, i.e. it depends on the relative size and position (because external pressure varies with position) of openings and the permeability of the cladding. Internal pressure to a first approximation can be calculated on the assumption that the volume of air entering the building must equal the volume escaping. The quantity of air Q flowing through an opening is proportional to the area and the square root of the pressure drop across it. Thus:

$$\Sigma a \left(p_e - p_i \right)^{0.5} = 0$$

where $p_e = C_{pe}q$, and $p_i = C_{pi}q$. The external pressure coefficients of all faces should first be established, and the equation then solved for C_{pi}.

Codes of practice generally provide values for C_{pi} and calculation is seldom undertaken.

Earthquake loading

Although earthquakes do not significantly affect the UK, it would be useful to have some appreciation of the nature of earthquake loadings. There are many other parts of the world subject to seismic forces (see Figure 5.15).

Sudden movements of the Earth's crust cause the Earth's surface to vibrate, and this in turn is transmitted to the structures via their foundations. The dynamic motion of the ground causes acceleration both in the horizontal and vertical direction. In general the horizontal component is of greater interest to structural engineers, although increasing attention is being shown to the vertical motion. The size of an earthquake is measured in two ways: with reference to the *energy* released by the geological disruption, and with respect to the *damage* caused. The first is the measure of the *magnitude* and the logarithmic Richter scale is used. There is no reference to the distance from the *epicentre* (the location where the body waves through the Earth's crust reach the surface). The damage caused on a site is referred to as the *intensity* of the earthquake, and the Modified Mercalli Intensity scale (MMI) or the Rossi-Forel ratings are used as standards for classification of damage.

Seismic maps quantifying the risk have been produced for different parts of the world. These are based on historical data of earthquakes and the location of geological faults. The maps usually show the intensities as peak ground acceleration (% *g* where *g* is the acceleration due to gravity) associated with particular probabilities of exceedence.

Consider a simple model of the structural behaviour during an earthquake. When the ground suddenly moves, it takes time for the upper part of any structure to catch up. The structure therefore takes up a deflected shape. We can calculate the equivalent static force that would cause this deflection, and clearly the building will be satisfactory if it can withstand this equivalent force.

Structural design is carried out using a full dynamic analysis for tall or important structures. However, a relatively simple equivalent static analysis is useful, particularly for preliminary design. It also highlights the various factors that influence seismic design. The procedure is discussed below.

During ground motion, the inertia of a building results in a horizontal *shearing* force, proportional to the weight supported at a given level and the acceleration of the ground motion. Thus shear $Q = C_s Mg$, where C_s is the seismic coefficient, M is the mass of the building above the level under consideration and g is the acceleration due to gravity. The seismic coefficient is usually expressed as a fraction of g and depends on the seismicity of the locality, local soil conditions, the natural period of vibration of the structure and the ductility of the structure (ability to absorb energy through deformation). One formulation to reflect these is:

$$C_s = ASI/R$$

where A is the site-dependent effective peak acceleration read from seismic zoning maps, S is the seismic response factor, which is a function of the period of vibration in the fundamental mode and the soil/characteristics. I is the 'importance' factor which takes into account the consequences of failure, and R is a factor intended to take into account the ability of the structure to absorb energy by ductile deformation without collapse. This implies some damage and the need for repair. Unlike the design for normal loads, it will be grossly uneconomic to design a structure to remain in the elastic range, i.e. with limited deformation under earthquake conditions. Most structural materials possess considerable reserves of strength after first yield, and seismic design taps the strength in the inelastic phase. The value of R is usually presented as a function of the structural system and the materials used.

Highway loading

The loadings to be used in the design of highway structures are generally specified by the appropriate authorities in each county. This is done by way of reference to a standard (BS 5400 in the UK) and additional requirements.

Probabilistic considerations are important in determination of traffic loading on bridges. Where the loaded length is sufficient to admit more than one vehicle, a governing condition will occur when traffic is brought to a standstill, minimising the separation between vehicles. Does one take a 'garage' of heavy vehicles over the whole bridge, or light vehicles, or a mix? Loading models are usually based on traffic flows and mixes recorded at various sites to represent conditions on the heavily trafficked commercial routes and extrapolated to take into account the predicted growth of motorway traffic.

In the UK, standard highway loading consists of HA and HB loading. *HA loading* is a formula loading that represents normal traffic conditions. *HB loading* is an abnormal vehicle unit loading and derives from the nature of exceptional industrial loads (e.g. electrical transformers, generators and pressure vessels) likely to use the road.

HA loading consists of either a uniformly distributed load and a knife-edge load, or a single wheel load. The carriageway is divided into notional lanes and the prescribed loading is applied to each lane. The uniformly distributed load is a function of the loaded length. Up to a loaded length of 30 metres the loading in BS 5400 represents closely-spaced vehicles of 25 tonnes laden weight in two lanes. For longer loaded lengths the spacing is progressively increased, and medium weight vehicles of 10 tonnes and 5 tonnes are interspersed.

The loads given in BS 5400 make allowances for impact effects caused by irregularities of the road surface and by imbalances in the vehicles.

In addition to traffic loads, other loads to be considered in the design of highway bridges include:

- wind loading (on the structure and the vehicles)
- centrifugal effects in curved bridges
- temperature effects (expansion/contraction and friction at bearings caused by the changes in mean temperature and effects within the structure caused by differential temperature between upper and lower surfaces)
- effects of shrinkage, creep and residual stresses
- differential settlement of supports
- longitudinal loading caused by traction on braking
- loads due to vehicle collision with parapets
- earth pressures on retaining abutment structures
- collision loading on supports over railways, canals or navigable water.

Codes of Practice prescribe the key parameters to be considered for each of the above. They also lay down rules for combining the various loadings.

Further reading

BS 5400–2 , *Steel, Concrete and Composite Bridges Part 2: Specification for Loads.*

BS 6367, *Code of Practice for Drainage of Roofs and Paved Areas.*

BS 6399–1, *Loading for Buildings Part 1: Code of Practice for Dead and Imposed Loads.*

BS 6399–2, *Loading for Buildings Part 2: Code of Practice for Wind Loads.*

BS 6399–3, *Loading for Buildings Part 3: Code of Practice for Imposed Roof Loads.*

BRE Digest 436 Parts 1, 2 and 3, *Brief Guidance for Using BS 6399–2:1997 and Worked Examples.*

BS 8100–1, *Lattice Towers and Masts Part 1: Code of Practice for Loading.*

Cook, N. J. (1985) *The Designers' Guide to Wind Loading of Building Structures Part 1,* Butterworths, London.

Currie, D. M. (1999) 'Changes to standards on imposed loads for buildings,' *The Structural Engineer* 77(4): 22–28.

Dickie, J. F. and Wanless, G. K. (1993) 'Spectator terrace barriers', *The Structural Engineer* 71(12).

ENV 1991–1, *European Pre-standard: Basis of Design and Actions on Structures.*

ENV 1995–1, *Design of Timber Structures.*

Mitchell, G. R. and Woodgate , R. W. (1971) *Floor Loadings in Office Buildings – The Results of a Survey,* Building Research Station Current Paper CP 3/71.

—— (1977) *Floor Loadings in Domestic Buildings – The Results of a Survey,* Building Research Station Current Paper CP2/77, 216–22.

—— (1977) *Floor Loading in Retail Premises – The Results of a Survey,* Building Research Station Current Paper CP 25/77.

Smith, J. W. (1988) *Vibration of Structures – Applications in Civil Engineering Design,* Chapman and Hall, London.

Chapter 6

Material properties

Introduction

Good design requires a proper appreciation of the properties of materials used in construction. While this is a vast specialist subject area for material technologists, some grounding in this area is essential for engineers.

The purpose of this chapter is to highlight some of the key properties of the more commonly-used construction materials. For a more exhaustive treatment, specialist literature should be consulted.

There are trade associations for most materials, such as the Brick Development Association, British Cement Association and Aluminium Federation. These bodies usually provide good advice and also produce helpful publications. A designer has to learn to search for information from these bodies and other sources such as BRE (Building Research Establishment), CIRIA (Construction Industry Research and Information Association) and BSI Codes and Standards.

It should also be borne in mind that new materials appear on the market all the time and an engineer should be capable of evaluating these critically. The information provided in this chapter, covering only a limited number of materials should induce curiosity and equip the designer to ask penetrating questions when confronted with new situations.

An engineer will be generally concerned with the following material properties: strength, deformation under load, performance in fire conditions and durability. In this chapter we shall discuss the above features of a selected number of materials. It will, however, be useful first to define some of the technical terms that will crop up and also to provide some general background.

Strength

The colloquial term *strength* is not sufficiently precise for design purposes. There are many types of strengths of interest to a designer. When a member is subject to an axial pull at each end, we say that it is subject to *tension* and at all cross sections there will be *tensile stress*. The member will not fail if these internal stresses are less than the tensile strength of the material. Similarly if the same member is pushed at both ends by equal and opposite forces, it is said to be under

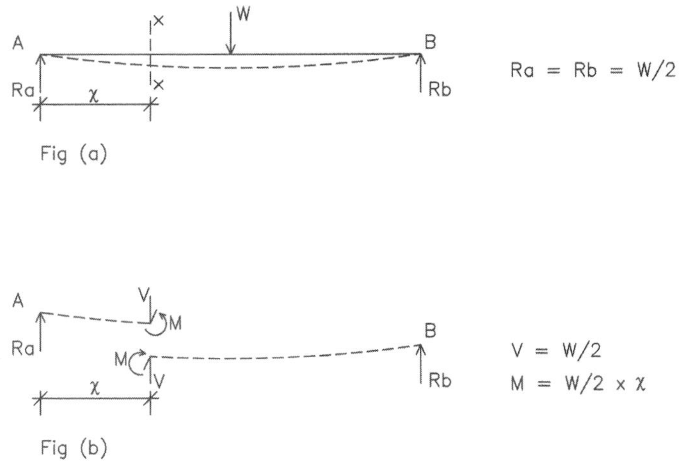

Fig (a)

$Ra = Rb = W/2$

Fig (b)

$V = W/2$

$M = W/2 \times \chi$

Figure 6.1 Member in bending: (a) external forces; (b) internal forces.

compression and the internal stresses at all cross sections are compressive. There is a corresponding compressive strength that must not be exceeded if failure is to be avoided. (Failure in compression is not straightforward as the member could fail by buckling, if it is too thin or slender. However, for the initial understanding, the above description should suffice.)

Tension and compression may be induced in a member even though it is not subject to axial forces as described above. Imagine a plank of wood spanning two supports. If a load is placed on it, it will bend. Such a member is called a flexural member. In this example the bottom surface will elongate and the top surface will shorten. Thus the top fibres will be in compression and the bottom fibres in tension. The flexural strength (tensile and compressive) could be different from strengths under axial loading.

Then there is the shear strength of materials. In the example quoted above, imagine that the load is at the centre. Under the action of the load the plank will bend, as shown in Figure 6.1(a). We can cut the plank at section XX and still preserve the exact bent shape, provided we supply the external forces V and M at the cut ends as shown in Figure 6.1(b). It can be seen that these forces are required to maintain the equilibrium of plank lengths AX and BX. The forces would have existed as internal forces before the plank was cut. The vertical internal force V at the section is referred to as the shear force, and it causes shear stresses in the plank. As long as these remain less than the shear strength of the material, shear failure will not occur. The magnitude of the stresses will generally vary depending on the location along the span and within the cross-section. In some cases special failure criteria are used to allow for the combined effects of shear and bending stresses.

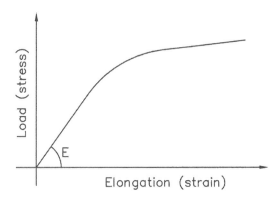

Figure 6.2 Typical load–elongation relationship.

Bond strength, as the name implies, describes the interface strength between two materials. Bond between reinforcement and concrete in reinforced concrete is an obvious example.

Strength can also be time-dependent. For example, concrete gains most of its strength over a period of 28 days.

Standard tests exist to evaluate most material properties.

Deformation under load

Consider a steel rod subject to axial tension in a testing machine. As the load is progressively increased, its length will increase. A plot of the load/elongation curve might be as shown in Figure 6.2.

By concentrating on the initial linear part of the plot, it will now be possible to predict the elongation or deformation at any load. While useful, this plot applies only to the particular bar. A more general representation will be in terms of stress and strain. This can be achieved as follows. The value of the ordinates divided by the cross-sectional area of the bar will represent the stress. Similarly, the elongation can be converted into strain by dividing the abscissa by the original length of the bar. We have now re-labelled the original plot in terms of stress and strain. For any stress (no longer dependent on the bar size) strains can be predicted. The slope of the linear portion of the plot is (stress ÷ strain). It is referred to as the *modulus of elasticity E*. Thus strain at a stress of σ is (σ/E). In turn, the actual deformation can be calculated if the length of the member is known.

The E value fo the member, calculated as described above, is used for axially-loaded members as well as flexural members (subject to bending). The value of E is an important property in the context of stiffness of structural members. The slope of the stress–strain diagram will be steeper for stiff materials (high E values) compared with less stiff materials. At this juncture it should be pointed out that some materials, e.g. concrete, exhibit a non-linear stress–strain relationship. However, this does not invalidate the concept, and particular methods exist to obtain the E values under these conditions.

The shape of the stress–strain curve beyond the linear stage depends on whether the material is ductile or brittle. Ductile materials exhibit a long plateau when the strain increases with very little increase in stress until a breaking value is reached. This is referred to as *plasticity*. Brittle materials reach the breaking strain with little plasticity.

The discussion above is concerned with the instantaneous response of a member to loads. Some materials, such as concrete or timber, exhibit a time-dependent response called *creep*. In other words, with no change in loading the deformation of a member increases in the long term above the instantaneous value. It is as if the material has softened and the slope of the stress–strain curve has become shallower compared with how it started life.

Steel cables under constant stress exhibit a phenomenon similar to creep called *relaxation*. A small percentage of the initial force is lost.

Performance under fire conditions

Fire is an important design situation and to ensure safety an understanding of the behaviour of materials in elevated temperatures is required.

One important strategy in fire-safety engineering is the provision of adequate fire resistance to structural elements, to safeguard the stability of the structure. Some materials (e.g. brickwork) possess good natural resistance, but others (e.g. steel) may require additional protection. The level of protection will depend on a number of factors such as the time required for occupants to escape, whether parts of the building have to act as refuges in fire situations, whether it is necessary for fire fighters to work inside the building to extinguish the fire, and any special insurance requirements to ensure the survival of the structure.

All materials lose strength with increase in temperature. Some idea of the temperatures likely to be encountered can be gained by reference to an internationally- accepted standard time–temperature curve (Figure 6.3).

The fire resistance period required for structural elements depends on a number of factors including the occupancy or use of the building (reflecting the fire load in the building), their location (height above or below ground, reflecting the degree of difficulty in fire fighting), and whether active fire protection (such as sprinklers) has been installed. Generally the required period is in the range of 30–240 minutes. Some limited numbers of structural members (e.g. members supporting only roofs) are exempt from fire resistance requirements.

So what are we looking for from a material in terms of fire resistance? Three criteria are used: load-bearing capacity (strength for the duration), integrity (resistance to thermal loads and to passage of smoke through cracks and gaps), and insulation (sufficiently low thermal conductivity to limit the temperature rise on the far side of the surface not subject to fire).

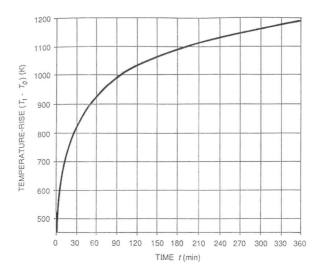

Figure 6.3 Standard curve of time v. temperature rise.

One way to satisfy these requirements is to expose the element in a furnace to the 'standard' fire, i.e. heating the furnace to the standard time–temperature curve. Criteria for evaluating the performance are laid down in national standards. This approach is expensive and is practical only for small mass-produced elements. Therefore other analytical and empirical methods have evolved.

Deterioration mechanisms

An implicit assumption in design is that the durability of a structure in its environment is maintained during its working life. This requires an appreciation of the performance of materials when they are subject to various environmental influences. These influences are mainly chemical or biological in nature.

Examples include corrosion of steel, carbonation of concrete, sulphate attack of concrete and cement mortars, rotting of timber and embrittlement of plastics and mastics by ultraviolet light. Understanding the mechanisms involved will help in the choice of appropriate materials (or compositions) or in taking appropriate protective measures. This is very much part of design as emphasised in Chapter 1.

Reinforced concrete

General background

Concrete is a versatile man-made material. Its main constituents are cement, coarse aggregate (stones), fine aggregate (sand) and water. Cement reacts chemically with water to produce a paste, which binds the matrix together. When mixed together the material is plastic; with time, it sets, hardens and gains strength. Until adequate strength is gained the wet mix is supported on, or contained in, shuttering. The mix can be poured or pumped. It can be moulded into different shapes, which ability distinguishes concrete from other materials. When wet concrete is placed in its final position in the structure, it is referred to, as concrete *in situ*. Alternatively, concrete elements can be made off-site in factories and then transported and assembled together on site. This is called *pre-cast concrete*.

Concrete is basically artificial stone and, as such, has the same basic properties as stone. Its great advantage is that, as a man-made material, it can be poured into moulds of any shape where it sets, so it is not necessary to form the material by carving, as is done with stone. A further advantage is that its properties may be tailored to a considerable degree to meet different situations.

The basic description of concrete, its strengths, advantages and weaknesses were discussed in Chapter 4, together with the reasons for reinforcing plain concrete.

Now the properties of concrete and reinforcing steel and how they may be controlled will be considered in more detail.

Constituent materials of concrete

Cement

There are many types of cement with different properties. The most commonly used cement for concrete production is *Ordinary Portland Cement* (OPC). The main chemical compounds in OPC are tri- and dicalcium silicates, tricalcium aluminates and tetracalcium aluminoferrites. When mixed with water, the hydration results in a crystal structure and calcium hydroxide (lime). Heat is also produced. *Rapid hardening Portland cement* is more finely ground and gains strength more rapidly as a result. *Low heat Portland cement* is less reactive than OPC, and heat is generated more slowly. *Sulphate-resisting Portland cement* has less tricalcium aluminate and is used when there is a risk of sulphate attack (see the Durability section, later). There are a number of composite cements notable among which are *Portland-blast furnace cement* and *Portland PFA (Pulverised Fuel Ash) cement*. Blast furnace slag and pulverised fuel ash are also added to concrete mixes as cement replacements. These materials arise as waste products in other industrial processes and, with the current emphasis on sustainability all around the world, are used widely as a waste minimisation strategy. Ground granulated blast furnace slags (ggbfs) are made by grinding the vitreous granular material produced by the rapid quenching of molten slags formed during the manufacture of pig iron. Ggbfs has cementing properties that are less reactive than OPC; it generates heat less quickly, and gains strength more slowly. It is also more resistant to sulphate attack than OPC. Pulverised Fuel Ash (PFA) is extracted from flue gases of furnaces that use bituminous coal. The rate of heat generated and the strength gain are slower than OPC. A PFA/OPC mix can be more chemically resistant.

Condensed silica fume is another cement replacement commonly used in the production of high-strength concrete and concretes for high chemical resistance. Silica fume is finer than cement and fills the interstices between cement particles, thus producing a very cohesive mix. PFA and silica fume are also referred to as *pozzolan*, which occurs naturally in Italy. Chemically, it is silica that reacts with the lime (calcium hydroxide) produced as a result of hydration of OPC and produces additional cementing properties and strength. Pozzolans were used extensively by the Romans to make concrete, and many such concrete buildings still exist. The Pantheon in Rome is possibly the most impressive example.

Aggregate

The aggregates normally used for concrete are natural materials: hard crushed rock or gravels. Recycling is gaining more importance in the context of sustainability and there are encouraging test results on the use of recycled aggregates made from old concrete. Natural sands and crushed rock sands are used as fine aggregates. Aggregates form a substantial volume of concrete, and their properties have a marked effect on the resulting concrete. Natural aggregates are used to provide 'normal weight' concrete, which has a density of around 2500 kg/m^3. Lightweight concretes with densities of around 1600 kg/m^3 are also available. These use aggregates made principally from sintered PFA or expanded clay, shale or slate. Lightweight concrete has better thermal insulation and fire resistance properties, but greater shrinkage and moisture movement.

For radiation shielding or ballasting, heavyweight concrete with density up to 3700 kg/m^3 is produced. These use heavy aggregates such as barytes, iron ores or iron shots.

Admixture

In addition to the principal constituents mentioned above, admixtures are sometimes used to confer particular properties. These include: workability aids, water-reducing admixtures (called plasticisers), air entraining agents (to improve resistance to freezing and thawing in outdoor construction), accelerators (to speed up the hardening of wet concrete, particularly in cold weather) and retarders (to delay the hardening process, which is useful when dealing with large volumes of concrete or when there is a risk of rapid setting in high temperature conditions).

Concrete mix design

Design of a mix to achieve particular properties in fresh and hardened concrete is an important aspect of concrete technology. There are many interacting parameters which come into play, and they include water/cement (w/c) ratio, maximum size of aggregate, grading of aggregates, aggregate/cement ratio and use of admixtures. Usually trial mixes form the basis for mix design. In the UK, most sites do

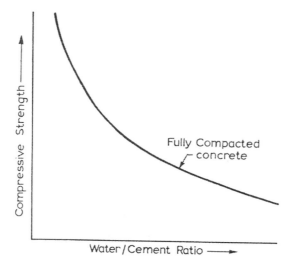

Figure 6.4 The relation between strength and water/cement ratio of concrete.

not now have their own concrete-making facility and it is usual to have concrete delivered from a ready-mix plant. The concrete supplier takes the responsibility for the mix design and the quality of concrete supplied to site. Designers specify their requirements and limits on some parameters (such as the 28-day strength, maximum w/c ratio, minimum and maximum cement content and size of aggregate) and rely on the substantial expertise of the ready-mix depots to achieve the desired properties economically.

The water content of a mix, usually measured in terms of water/cement ratio, virtually decides many of the important properties of concrete such as density, strength, durability and permeability. A typical relationship between strength and w/c ratio is shown in Figure 6.4. From many considerations, the mix designer aims for as low a value of w/c ratio as possible, i.e. for a given cement content there should be just enough water to achieve full hydration. Any excess water simply forms voids and adversely affects many of the properties mentioned above. But will such a mix be sufficiently workable? In other words, will it be fluid enough to allow it to be compacted into the cohesive matrix which is essential to achieve the density, strength and durability? Workability interacts heavily with w/c ratio.

The size of the aggregate is governed by the dimensions of the concrete member. In general a maximum size of 20 mm is common; for narrow or thin members 10 mm aggregate (which is more expensive) is specified. It should be noted that while the maximum size of the aggregate is specified, the aggregate is not supplied in a single size generally. It is graded, and national standards specify the proportions of various sizes (*grading limits*) of aggregates to achieve cohesive concrete. The grading of the aggregates will have a bearing on the cement content as the cement fills the voids between the aggregate particles.

Properties of concrete

The compressive strength of concrete at 28 days is almost universally adopted as the basis for structural calculations. It is also used to designate concrete. The strength gain after 28 days is generally not significant. When concrete is placed on site, samples of concrete are taken to make cubes (or cylinders) that are cured in a specified manner and tested at different ages. In the UK cube strength is used, although design information in Eurocode 2 for concrete is in terms of cylinder strength. The range of concrete commonly used is 20 N/mm^2 – 50 N/mm^2 (cube strength). Currently, there is a lot of interest in high-strength concrete (cube strength of about 90 N/mm^2) particularly for applications such as offshore platforms.

Strength development with age is not linear. A typical relationship is shown in Figure 6.5 for concrete with OPC. Mixes with pfa and ggbfs gain strength more slowly. This has implications for striking the shuttering. The strength development is also affected by ambient temperature. Figure 6.6 shows a plot of strength development against temperature.

Strength properties

Properties noted in this section are in accordance with Eurocode 2, in which the 28-day cylinder strength is used as the basis. First a brief word on the need for a statistical approach. Even when concrete is produced from nominally the same mix proportions, variation in properties will occur with time. For instance, if the compressive strength of concrete supplied to a site is measured throughout the duration of a project, it will be found that there are variations, see Figure 6.7. The variations follow a statistical distribution called *normal distribution*. In design, the *characteristic* strength, rather than the mean strength, is used. This strength is defined as the level below which only a small proportion of all results are likely to fall. It is customary to use a five per cent value as the characteristic value, and this is represented by f_{ck} (the suffix k denotes the characteristic value). Likewise, when concrete is ordered, it is a concrete with some specified characteristic strength that will be asked for.

Whether the concrete supplied to a site complies with the specified strength is checked on a statistical basis to ensure that a strength less than the characteristic has only a five per cent chance of being received. To ensure this, the producer has to provide concrete with an average strength considerably above the specified characteristic value. The amount by which the average exceeds the characteristic will depend on the effectiveness of the producer's control measures.

Codes of Practice relate most properties to f_{ck}. The following are in accordance with Eurocode 2 for concrete strengths up to 50 N/mm^2.

Mean value of compressive strength	$f_{cm} = (f_{ck} + 8)$ N/mm^2
Mean value of tensile strength	$f_{ct,m} = 0.3 f_{ck}^{0.67}$ N/mm^2
5% Characteristic value of tensile strength	$f_{ct,k, 0.05} = 0.7 f_{ct,m}$ N/mm^2
Mean E value	$E_{c,m} = 9.5 f_{c,m}^{0.33}$ N/mm^2

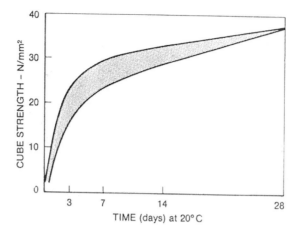

Figure 6.5 Envelope of strength development for a number of OPC concretes.

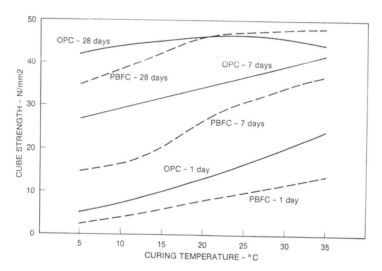

Figure 6.6 Effect of curing temperature on the strength gain of an OPC and PBFC concrete. OPC is Ordinary Portland Cement, PBFC is Portland Blast Furnace Cement.

(As the stress-strain diagram is not linear, the slope of a line joining the origin and the point representing $0.4 f_c$ is taken as the E value; f_c is the maximum stress).

Strain at compression failure $\qquad\qquad \varepsilon_{c,k.} = 0.0035$

Characteristic value of shear strength $\qquad \tau_{R,k} = 0.25 f_{ct,\,k,0.05}$ N/mm^2

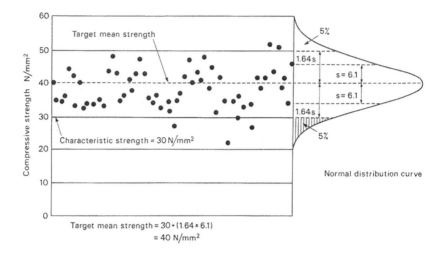

Figure 6.7 Variation of concrete strength on one site.
Target mean strength = 30 + (1.64 × 6.1) = 40 N/mm².

Characteristic value of bond strength:

High bond (ribbed) bars	$f_{b,k} = 2.25 f_{ct,k,0.05}$ N/mm^2
Plain bars	$f_{b,k} = 0.36\sqrt{f_{ck}}$ N/mm^2

Reinforcement of concrete

Low carbon steel is the most commonly used as reinforcement for concrete. The surface of the reinforcement may be plain, ribbed or indented. The last two are called *high bond* bars. In Europe, the trend is to supply only high bond bars. The reinforcement may be delivered as loose bars or as welded mesh or prefabricated cages for each element.

The yield strength f_y of reinforcement is used in design. It is assumed to have the same value both in tension and compression and is the stress when the linear stress–strain relationship ends. Typical characteristic value $f_{y,k}$ used in the UK is 250 N/mm^2 for plain bars and 460 N/mm^2 for high bond bars. There is a move to increase this value to 500 N/mm^2. The E value for all reinforcement is taken as 200 × 10^3 N/mm^2. *Ductility* is an important property of reinforcement and requirements are laid down in the relevant standards. The main reasons for requiring a ductile performance of the structure include:

- the need to withstand accidents without collapse
- to provide warning of incipient collapse by development of large deformation prior to collapse

- to be able to redistribute the stresses within the structure from highly stressed locations
- to absorb energy during seismic loading by large deformation without collapse.

Most of these attributes are conferred by the ductility of the reinforcement.

Durability of reinforced concrete

To achieve durable concrete requires an understanding of how concrete deteriorates in various environments. The main mechanisms of deterioration are *carbonation chloride attack, sulphate attack, thermasite attack, alkali silica reaction* and *frost attack*. Carbonation and chlorides do not harm the concrete, but they provide the conditions necessary for the reinforcement to corrode. The various mechanisms will be discussed briefly.

Carbonation

Carbonation is an unavoidable chemical reaction in all concrete exposed to air. Atmospheric carbon dioxide combines with calcium hydroxide in the hardened concrete to produce calcium carbonate. The significance of this reaction lies in the fact that the highly alkaline environment of the fresh concrete, which protects any embedded metal (reinforcement) from corrosion, is gradually eroded by the carbonation process. The rate at which the carbonated front advances into the concrete depends on many factors, including the permeability of the concrete, the extent and depth of cracking, the humidity (50–75% RH produces the highest rate) and the concentration of acid gases in the atmosphere. For well-made dense concrete, the rate will be so slow that the cover concrete prescribed in national standards will retain sufficient alkalinity to protect the reinforcement for the life of the structure. Ensuring a good quality concrete cover to reinforcement bars is the best way of avoiding corrosion of the reinforcement, which is highly disruptive and expensive to repair.

Chlorides

Another cause of corrosion of reinforcement is chloride attack. When present in sufficient quantities, chloride ions promote corrosion even in highly alkaline conditions. The mechanism is rather complex but it is thought that there is an increase in electrical conductivity of the pore water and as a result of its electro-chemical properties, corrosion is induced. Sources of chlorides could include accelerators used as setting agents in cold weather, chlorides in aggregates (marine origin), use of de-icing salts in winter, and exposure of the structure to marine environment. Codes of practice set strict limits on the amount of chlorides in aggregates and also prescribe high quality concrete and increased cover to reinforcement.

Sulphates

Sulphates in aqueous solution attack the cement in hardened concrete and produce an expansive reaction that depends on the type of cement and aggregate. Calcareous binders and limestone aggregates are soluble over time in acidic water. Critical factors affecting the rate of attack include the concentration and type of sulphate and the pH in soils or ground water, the water table and mobility of ground water, quality of concrete (particularly its compaction and permeability) and presence of frost. Sulphates occur naturally in many soils, such as London clay, Lower Lias, Oxford clay, Kimmeridge clay and Kuper marl. Sulphides can occur naturally (pyrites) or from industrial wastes. Some materials used as fill below ground floors (colliery shale, bricks, etc.) could also act as sources of sulphates. The considerable expansion of the concrete caused by sulphate attack will lead to its final breakdown. The soil and ground water are tested for the presence of sulphates and are classified with respect to the concentration found. Codes of Practice include advice regarding the type and quality of cement that should be used for each class. Cement with low tricalcium aluminate content provides excellent resistance to sulphate attack. Good sulphate resisting properties also exist in concretes using OPC in conjunction with pfa (25–40%) or ggbfs (70–90%).

Thaumasite attack

The presence of limestone aggregates in concrete can increase its overall vulnerability to attack by acid. In cold (less than 15°C) and very wet conditions, it has been found that concrete containing calcium silicate (from the cement itself), sulphate (from an external source such as ground or water), carbonate (limestone fillers in cement, limestone aggregate) and reactive alumina (from cement) suffers a particular form of sulphate attack, resulting in a mineral called *thaumasite*. This is also an expansive reaction leading to breakdown of concrete. Use of sulphate-resisting cement in these conditions is not always an effective means of ensuring durability. According to current thinking, the method of avoiding thaumasite attack is not to use ground limestone and limestone aggregates in concrete that is subject to temperatures below 15°C and consistently high relative humidity.

Alkali silica reaction

Alkali silica reaction is a mechanism of deterioration that results from an interaction between alkaline pore fluids in concrete and certain types of aggregates. The pre-conditions are a sufficiently alkaline solution, a susceptible aggregate and sufficient supply of water. Siliceous aggregates are the most common type of aggregate susceptible to alkali attack. The reactivity of the different silica minerals depends on the amount of order in the crystal structure. Opal is the most reactive and strained quartz is normally unreactive. There are other types of silica of intermediate reactivity (glass, micro-crystalline quartz, chalcedony, etc). Reactive minerals have been found in some sand and gravel dredged off the south coast, Bristol Channel and Thames Estuary and in some land-based quarries in south-west England and the Trent and Thames valleys.

The concentration of alkalis in cement is expressed as equivalent sodium oxide. Experiments with aggregates containing an opaline constituent (the most reactive) have suggested a limit of three kilograms equivalent sodium oxide per cubic metre of concrete to prevent the reaction.

The structural effects of alkali silica reaction include surface cracking at an early stage, marked reduction in tensile strength and elastic modulus and bowing caused by expansive pressures.

The expansion produced by the reaction can be reduced to acceptable levels by the use of cement replacement materials such as natural pozzolan, pfa and ggbfs. Low alkali OPC is also effective.

Frost attack

Poor concrete suffers when subject to cycles of *freezing and thawing* (commonly called frost attack). Good quality concrete is resistant to frost unless de-icing salts are used. Air entrainment is usually specified to improve the durability of concrete in outdoor conditions, subject to freezing and thawing.

Summary

From the above discussion it should be obvious that well-compacted, cohesive, dense concrete is a fundamental requirement for durability in all circumstances. The constituent materials and mix proportions should be carefully monitored for each environment. At a practical level Codes of Practice define exposure classes based on different deterioration mechanisms, and recommend minimum cement content, maximum water/cement ratio and minimum cover to reinforcement. All these requirements should be met simultaneously. Durability considerations will be the starting point in design and the strength of the resulting concrete mix is then used for the calculation of mechanical resistance.

Fire performance of concrete

Concrete structures possess excellent natural fire resistance. Concrete encasement of steel structures used to be a common method of fire protection. The fire performance of lightweight concrete is superior to that of normal-weight concrete. In practice, the fire resistance of individual elements is considered in isolation although it is generally better to consider the resistance of the whole structure. The study of the latter is relatively more complicated.

Like all materials, concrete and reinforcement lose strength at elevated temperatures. Figures 6.8 and 6.9 show, for various temperatures, the reduction in the strength of concrete and reinforcement as a fraction of the strength at 20°C. It will

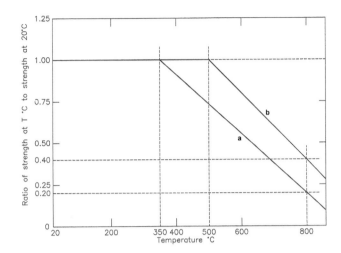

Figure 6.8 Design curves for variation of concrete strength with temperature: (a) dense concrete, (b) lightweight concrete.

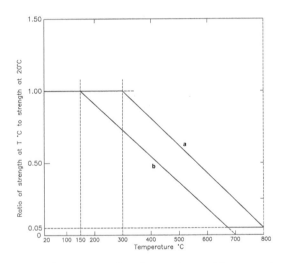

Figure 6.9 Design curves for variation of steel strength or yield stress with temperature: (a) high yield reinforcement bars, mild steel reinforcement bars and high strength alloy steel bars, (b) prestressing wires or strands.

be seen that concrete with lightweight aggregates retains a higher percentage of its strength. Cold worked reinforcement loses more strength than hot rolled bars. The strength of normal-weight concrete drops by fifty per cent at temperatures of about 600°C. Similarly the reinforcement strength drops by fifty per cent at a temperature of around 550°C. In the standard time–temperature curve, these temperatures can be seen to be reached in about 10 minutes.

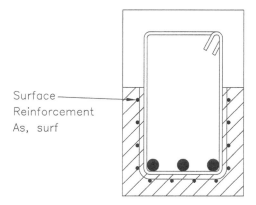

Figure 6.10 Surface reinforcement.

When heated to high temperatures, *spalling* of the outer layers of concrete occurs in some circumstances. Spalling is the result of rapid rates of heating, large compression stresses in the member and high moisture content in the member. Members with concrete cover in excess of about 40–50 mm are vulnerable. Spalling is undesirable as it exposes the interior of the concrete to high temperatures and consequently weakens it. Protection against spalling can be achieved in a number of ways, including applied finishes, use of lightweight or limestone aggregates and provision of surface reinforcement (Figure 6.10).

The insulation qualities of concrete can be studied by considering the temperature contours through concrete members. See Figure 6.11 for slabs heated from the soffit and Figure 6.12 for beams heated on three sides.

The fire resistance of structural members is affected by a number of factors including their size and shape, the disposition and properties of reinforcement, the load supported in fire conditions, the type of concrete and aggregate, protective cover to reinforcement and boundary conditions of the member, such as continuity (which provides an opportunity to shed loads to adjacent members).

There are three methods for assessing fire resistance of reinforced concrete members and these are tabulated data, fire tests and fire engineering calculations.

Codes of Practice provide information on the minimum dimensions (to provide the necessary insulation to the far side) and concrete cover thickness (to ensure that the reinforcement retains at least 50% of its normal strength). Information is provided for both normal-weight and lightweight concrete and for slabs, beams, columns and walls. The information is based on test results.

Fire tests are not normally undertaken other than for mass-produced members such as pre-cast slabs. It is not a practical solution for other structural elements.

Fire engineering calculations are now possible for some phenomena, such as flexural design. Data is still limited for checking shear, bond and anchorage. The flexural design starts with a prediction of temperature distribution through the

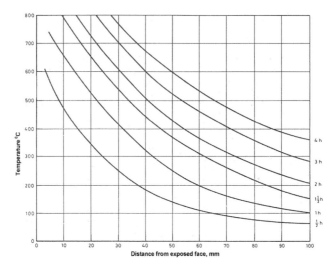

Figure 6.11 Temperature distribution in a dense concrete slab.

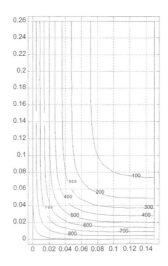

Figure 6.12 Temperature distribution in a dense concrete beam for a 1 hour fire.

member for a given fire rating, which fixes the surface temperature on the heated surface. Then the corresponding mechanical strength of the concrete and reinforcement can be derived using the reduction factors such as those shown in Figures 6.8 and 6.9. From this data it is a relatively easy matter to arrive at the moment of resistance of the section.

Summary

Concrete structures possess excellent fire resistance. Lightweight concrete fares better than normal-weight concrete. Members have to have a minimum dimension, which will provide sufficient insulation to the unheated face. They also need to have sufficient concrete cover to the reinforcement to ensure that a reasonable proportion of its cold strength is available during fire conditions. Excessive cover could lead to spalling which should be prevented. Detailing the reinforcement with adequate structural continuity substantially enhances the fire resistance.

Structural steel

General background

In the UK, steel had been used in construction since about the end of the nineteenth century. Cast iron and wrought iron were used extensively in the eighteenth and nineteenth centuries. Currently, steel is an extremely popular material. The percentage of impurities (principally carbon) distinguishes the three metals. The carbon content differentiates the metallurgy of the three materials and has a significant effect on their properties. Carbon content in cast iron is typically around 4.5 per cent and it is negligible in wrought iron (≤ 0.25 per cent). The carbon content of steel is controlled to be less than 1 per cent.

Many alloy steels are also produced. In an alloy steel, percentages of special elements are added to carbon steels to produce particular characteristics. The most well-known alloy steel used in construction is stainless steel, which is produced in two generic types: *austenitic* and *duplex*. Austenitic stainless steels have 17–18 per cent of chromium and 9–12 per cent of nickel. They possess a good range of corrosion resistance. Duplex stainless steels have higher strength and wear resistance than austenitic grades.

Structural steel is produced in different profiles for use in construction e.g. beams (I profile), columns (H shapes) channels (⊔), angles (L), tees (T), hollow sections (circular, rectangular or square tubes) and many light section profiles. In most cases section sizes are standardised. In many cases sections with a nominal serial size are produced with different thickness of flanges and webs so that they have different geometric properties and different weights. This enables reasonably economic use of the material. When sections larger than the standard profiles are required, these are made up by connecting plates together, either by welding or bolting (e.g. plate girders).

A number of methods are available for turning steel into the range of products mentioned above.

- *Hot rolling* is the most extensively-used method. It involves heating the steel from steel-making plants (usually delivered in different sizes and referred to as ingots, slabs, blooms or billets) about 1300°C, passing the material between a series of rolls to reduce the thickness of the material, and converting it to various shapes.

- *Forging,* where the heated ingot is carefully worked with hammers or a press, is usually used for producing heavy mechanical components.

- *Extrusion* involves forcing the heated metal through a die of the required shape, and is generally used for producing special profiles.

- *Tube forming* is carried out by hot-working, cold-working or welding.

- *Cold rolling* involves reducing the thickness of unheated material by rolling pressure. The main products are flat bars and sheets.

- *Cold forming* is used to produce complicated shapes that are too thin for hot rolling, e.g. purlins and sheeting rails.

Jointing

The previous paragraphs describe how discrete components or elements of a structure are manufactured. After necessary fabrication, these are delivered to site, in lengths to suit the layout of the structural frame and taking into account limitations on length imposed by transport facilities (usually 12 m) and the maximum weight of components that could be handled on site. The total structure, which is a three-dimensional entity, needs to be created from the individual linear members, using various jointing techniques. Joints need careful design, taking into account the forces transmitted through them.

The most common form of jointing is *bolting*. In the UK bolts are ordinarily supplied in two strength grades. These are fitted into clearance holes, usually 2 mm larger than the bolt diameter (a bolt of diameter d mm cannot be fitted into a hole of diameter d). The clearance causes some flexibility in the joint since some slip needs to occur to allow the bolt to bear against the sides of the holes. When such flexibility is not acceptable, high-strength friction grip bolts are employed. These resist forces by friction between the plies at the interfaces.

The other common method of connection is *welding*. The joint is made by fusing the metal along the line of the joint. The melted metal from each component of the joint unites in a molten pool bridging the junction. A solid bond is formed with the parent metal as the joint cools and continuity of the material through the joint is established. The heat input for welding is usually through an electric arc, which operates between the end of an electrode (a steel rod) and the workpiece. The depth of the arc melt into the metal (the depth of penetration) is proportional to the intensity of electric current. The depth of penetration that can be achieved in manual welding is usually around 2 mm. Therefore, to achieve continuity through the whole thickness of the jointed elements, the edges of the element need to be prepared by being cut back along the joint. The groove so formed is then filled with metal melted from the electrode.

Welding is a skilled operation and is best done at works rather than on site.

Properties of structural steel

Strength

A typical stress–strain diagram for mild steel (Grade 43) is shown schematically in Figure 6.13. Grade 43 and 50 steels are most common. These grade designations are in accordance with BS 4360, and have been redesignated in the new Euronorms (ENs) as S 275 and S 355, where the numerical part of the designation refers to the yield strength (e.g. 275 N/mm^2). There are particular requirements where plastic design is to be undertaken. Design assumptions include that a rectangular stress distribution is achieved over the full cross-section (full yield), and that such a section is capable of plastic rotation at a constant moment until plastic hinges (sections with full yield) can develop elsewhere in the structure. The conditions that should be satisfied to realise these assumptions are stated as follows in BS 5950.

- The yield plateau should extend for at least 6 × the strain at yield.
- The ratio of ultimate strength/yield strength should be greater than 1.2.
- The minimum elongation must be 15% on a gauge length of $5.65\sqrt{S_o}$, where S_o is the original cross-sectional area of the test piece.

Steels produced in accordance with BS 4360 satisfy these requirements.

Other properties

The following are normally used:

- Modulus of elasticity $\qquad\qquad E = 205 \times 10^3\,\text{N}\,/\,\text{mm}^2$
- Poisson's ratio $\qquad\qquad\qquad v = 0.3$
- Coefficient of thermal expansion $\quad \alpha = 12 \times 10^{-6}\,\text{per}\,^\circ\text{C}$

Brittle fracture

It is known that steels may show great weakness when subject to shock, even though ordinary tension tests may appear quite normal. This resistance is also temperature related and there is a significant reduction at low temperatures. The resistance is measured in a notched bar test, in which the energy expended to break the specimen is established. The two common tests are the *Charpy* test and the *Izod* test. Codes of Practice dealing with the specification for the material prescribe the minimum values for the energy absorbed. Brittle fracture is considered only for locations where tensile stresses arise in service conditions. It will be particularly critical where the structure is exposed to low temperatures.

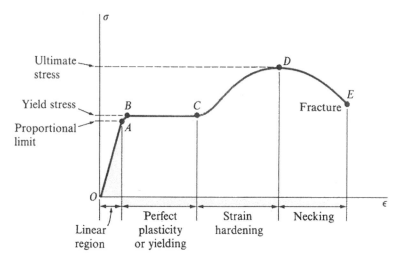

Figure 6.13 Typical stress–strain diagram for steel.

Hardness

Hardness is an essential quality for products such as armour plates, steel rails and tyres. Hardness is tested using either *indentation* tests or *scratch and abrasion* tests. The results are expressed as a *hardness number*, e.g. Brinell hardness number. Hardness tests are used to investigate the effect of heat treatment, hardening and tempering, etc.

There seems to be some correlation between the hardness number and tensile properties. Therefore some indication of the latter can be obtained from these tests. This will be useful in the context of investigation of fire-damaged structures, particularly the strength of connectors such as bolts.

Durability of structural steel

The main problem affecting the durability of carbon steels is corrosion, which is an electrochemical process. *Anodic* and *cathodic* zones form on the surface of the steel and, in the presence of moisture (electrolyte) and oxygen, an electric current is generated and the anode is dissolved. Areas that are readily exposed to atmospheric oxygen will be cathodic, while areas covered in dirt or rust scales will be anodic. These zones will shift and change as the corrosion reactions proceed. The product of the reaction is hydrated ferrous oxide, commonly called *rust*.

From the above brief description, it should be clear that both moisture and oxygen are essential for corrosion to take place. The rate of corrosion will depend on the impurities in the water. The conductivity of water increases with the amount of impurities in it. Thus in relatively clean environments such as rural areas, corrosion rates will be low; in polluted atmospheres such as industrial areas, the rate of corrosion will be higher. Marine environments, in which chloride ions are present, are particularly aggressive.

It follows from the above that the bare steel should be protected from moisture and oxygen to minimise the risk of corrosion. Various effective paint systems exist. The paint systems are themselves permeable and thus the thickness of the system is chosen to reflect the frequency of maintenance that is acceptable. Clearly the environment also has a direct effect on the thickness of protection, to achieve a given frequency of maintenance. The success of the protection also depends crucially on the surface preparation of the steel. The tolerance of the various paint types to the quality of preparation varies.

The detailing of the steelwork could also affect its durability. Details, which are likely to trap moisture, should be avoided as far as possible. For this reason in outdoor exposure, intermittent welds should be avoided and continuous weld should be used. Weepholes should be provided to let out any collected moisture.

Steels can be chemically modified to improve their corrosion resistance, for example in *weathering steel* elements such as copper, nickel and chromium are added. The structure of the rust layer is modified to produce a more finely-grained rust layer. Then there is stainless steel, which has been mentioned already.

Bimetallic corrosion (galvanic corrosion)

This is a reaction that occurs when two different metals are in electrical contact and are bridged by an electrolyte (normally water). Remembering the electrochemical reaction (which is the corrosion process), the baser metal becomes anodic and the nobler metal cathodic. The nobler metal tends to be protected at the expense of the base material.

The basic method of preventing bimetallic corrosion involves breaking the metallic path by insulating the dissimilar materials from each other, or breaking the electrolyte path by preventing the formation of a continuous bridge of electrolyte solution between the two metals. The former requires fitting insulating bushes and washers and the latter involves painting.

Fire performance of structural steel

Like all materials, structural steel loses strength with increase in temperature. This reduction starts at around 300°C. Melting temperature is approximately 1500°C. Stress–strain data for Grade 43A steel at various temperatures are shown in Figure 6.14.

In design, the load-bearing capacity at very high temperature is required to be in excess of the effects of loads applied under fire conditions. The temperature of the steel is thus the most critical parameter. All fire protection methods aim at keeping the temperature at a level commensurate with the required resistance.

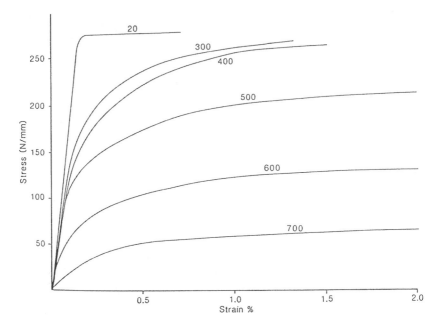

Figure 6.14 Stress–strain relationship for mild steel at different temperatures.

While provision for fire protection of steel is common in practice, modern fire engineering methods have demonstrated that unprotected steel can safely support the imposed loads in many circumstances. The most common among these is the justification of composite metal decks in which profiled steel decking acts not only as the permanent shuttering to the concrete infill but also as the tension reinforcement to the concrete under normal conditions of use. In fire conditions the metal decking is totally discounted, and the concrete infill with mesh reinforcement is justified by calculations that take into account the temperature gradients through the slab. Fire tests have confirmed the validity of this approach.

When required, fire protection can take the form of an applied coating of insulation, concrete encasement or hollow protection using boards.

Masonry

General background

The term masonry covers many types of constructions including *brickwork*, *blockwork* and *stonework*. In all cases masonry units (bricks, blocks or stones) are held together, generally with mortar. Some of the oldest structures in the world, e.g. the Pyramids and the Colosseum, are masonry. Applications of masonry include buildings, retaining structures and paving.

The compressive strength of masonry is a function of the strengths of the units and mortar. Strength is also derived through *bonding*, which describes the disposition of masonry units in a completed walling. There are many types of bonds, including English, Flemish and Quetta. Each bond has a characteristic arrangement of headers and stretchers (short and long faces of the masonry units) in each course (layer) of the masonry. It is not difficult to imagine that a wall in which the joints between the masonry units are staggered (in three dimensions) will be stronger than one in which planes of weakness are created through coincident joints. Some bonds, such as Quetta bonds, create hollows that permit the wall to be reinforced to resist lateral forces. (The Quetta bond evolved as a result of a major earthquake in Quetta in north-west Pakistan.)

All masonry is inherently stronger in compression than in tension. The strength of the masonry units contributes to the compression strength, but the tensile strength of masonry is largely determined by the strength of the mortar and the bond between the mortar and the units. Resistance to lateral loads is governed by the tensile strength of masonry, which is generally poor. It can be enhanced by a number of methods, such as the geometrical form (enhancement of section modulus) reinforcing the masonry (similar to reinforced concrete) and pre-stressing the masonry (inducing pre-compression).

Masonry may be used as cladding in framed construction or it can be the load-bearing element of the structure. Whatever form it takes, more than in any other material, the success of masonry depends on good design, careful detailing and sound workmanship on site. Design includes the selection of suitable materials, consideration of strength and stability, detailing for exclusion of moisture and movements, and consideration of thermal, acoustic and fire requirements.

Materials used in masonry

Clay bricks

This is the most commonly used masonry unit in the UK. The clay is fired in kilns at temperatures in excess of 1000°C to achieve ceramic bonding. The standard British brick format is 215 mm × 102 mm × 65 mm. Bricks are generally made with a *frog* which is a purpose-made indentation on either one or both bed faces of the brick. Most specifications require bricks to be laid 'frog up' and the frogs to be filled with mortar. Generally the bricks are solid although vertically perforated bricks are also produced.

Facing bricks are produced in various colours and textures and are used in walls without applying any finishes. The characteristic strength of bricks is usually in the range 7–100 N/mm^2 although commonly-used bricks are in the range 14–50 N/mm^2. The coefficient of thermal expansion is in the range $5–8 \times 10^{-6}$/°C. All fired clay units undergo an irreversible expansion as a result of adsorption of moisture from the atmosphere. The exact amount depends on the type of clay and firing temperature and it is typically in the range 0.02–0.12 per cent.

Calcium silicate bricks

These consist of a mixture of sand or crushed or uncrushed siliceous gravel. or a combination of these, with a lesser proportion of lime, mechanically pressed and the materials chemically bonded by the action of steam under pressure. Where only natural sand is used with lime they are referred to as sand lime bricks. Flint lime bricks contain a substantial proportion of crushed flint. The normal range of characteristic strength of these bricks is 14–48 N/mm². The movements of these bricks are higher than those of clay bricks. The coefficient of thermal expansion is $8{-}14 \times 10^{-6}/°C$. Drying shrinkage is 0.01–0.04 per cent. The bricks are normally produced to a modular format of 190 mm × 90 mm × 65 mm.

Concrete blocks

Dense concrete blocks are produced using cement and aggregates and are moulded under pressure and vibration. The length is in the range of 390–590 mm, the height 140–90 mm and thickness 60–250 mm. Commonly used blocks are 440 mm × 215 mm × different thicknesses. The strength is in the range 7–40 N/mm². Blocks can be solid or hollow. The coefficient of thermal expansion is $6{-}12 \times 10^{-6}/°C$ and the shrinkage coefficient is 0.02–0.06 per cent.

Autoclaved aerated concrete blocks

All the masonrymaterials described so far are generally used for their strength or appearance, or both. However, they are poor thermal insulants. In practice, the insulation is inversely related to strength. Strong units require a highly cohesive matrix of material with little voidage. Thermal insulation increases considerably if the structure of the material is cellular.

Autoclaved aerated concrete blocks are masonry products with a higher thermal insulation value. An aerating agent (usually aluminium powder) is added to a slurry of cement, pfa, lime and ground sand, then moulded as large blocks in autoclaves. Subsequently, they are cut to the required formats. The blocks are produced only as solid units. The normal strength of the blocks is around 3 N/mm², but they can be produced up to a strength of 7 N/mm² with good thermal insulation characteristics. The coefficient of thermal expansion is about $8 \times 10^{-6}/°C$. The drying shrinkage value is in the range 0.02–0.09%. The normal format for the blocks is 440 mm (long) × 215 mm (high) × various thicknesses.

Stones

Building stones are natural materials and are used exactly as they are found. Before quarried stone is used, seasoning and shaping will be required. Unlike bricks and blocks there is no processing or manufacture involved. Commonly-used building stones in the UK are limestones, sandstones, slates and granite. Strength is seldom a problem for normal usage. However, high-stress concentrations caused, for example, by differential settlements, restrained thermal expansion or corrosion of iron cramps, can seriously damage stones. Stones can be ordered in different sizes depending on the type of walling being considered. For ashlar masonry, in which units will be precise, consistent dimensions are required. Then there is random rubble masonry, which may or may not be coursed.

Mortars

Mortar is a key component of masonry required to join the units together and produce a strong composite unit. Mortars for masonry are based on building sands (soft sand) mixed with a binder such as cement and/or lime. Freshly-made mortar should be cohesive, i.e. it should hang on the trowel. It should also spread easily. It should not lose water readily, as this would cause it to stiffen immediately when it comes into contact with absorptive masonry units. Mortars should also remain plastic long enough to allow bricks or blocks to be adjusted easily to line and level.

In the UK, mortars use designated (i) to (iv). The new European Standards designate the same mortars as M10, M5, M2.5 and M1 to show the 28-day strength of the mortar in N/mm^2. The typical mortar compositions are shown in Table 6.1.

As the mortar designation increases from (i) to (iv) the strength of the mortar decreases but the ability of the mortar to accommodate movement increases.

Cement used for mortars is generally Ordinary Portland Cement. When there is a risk of sulphate attack, sulphate-resisting cement is used. The most likely source of sulphate attack is some types of clay bricks, or contact between masonry and sulphate-bearing ground such as retaining walls. The particular properties of mortar needed for laying masonry have led to the development of a variety of cements designed especially for use in mortars, and these are marketed as masonry cements. In the UK this type of mortar cement consists of a mixture of Portland Cement with very fine mineral filler and an air-entraining agent.

Mortars may be mixed on site or may be delivered to site as a ready-to-use factory-made product. The latter usually incorporates a retarder to delay the setting of the cement for up to 48 hours. Mortars can be coloured by the addition of suitable pigments.

Specifying materials

All types of masonry units are covered by British Standards and should be specified with reference to these. They cover such things as sizes, tests for durability, compressive strength, water absorption, efflorescence (for clay bricks), drying shrinkage (for concrete blocks), and acceptance/designation criteria.

BS 3921 covers the requirements for clay bricks and BS 6073 those for concrete blocks. BS 5390 is the Code of Practice for stone masonry and BS 8298 deals with

Table 6.1 Typical mortar compositions

Mortar designation	Volumetric proportions		
	Cement	*Lime*	*Sand*
(i)	1	0.25	3
(ii)	1	0.5	4–4.5
(iii)	1	1	5–6
(iv)	1	2	8–9

stone masonry in cladding. For detailed information, reference should be made to these documents; however, one or two important matters are noted below:

BS 3921 classifies bricks with respect to frost resistance as *frost resistant* (F), *moderately frost resistant* (M) and *not frost resistant* (O). Similarly, it distinguishes between two classes for soluble salt content: *low* (L) and *normal* (N). Limits are prescribed for various types of salts for Class L bricks. Thus there is a matrix of six possible designations for bricks, for example FL brick has frost resistance and low soluble salt content, and MN brick will be moderately frost resistant with normal soluble salt content. BS 5628: Pt 3 provides guidance on the selection of bricks (and blocks) for use in a variety of environments.

External stone masonry is selected in close consultation with the quarry concerned, based on past experience. There are no simple tests for determining the durability of limestones, which are principally affected by the porosity of the stone. The Building Research Establishment has published a guide (BRE Digest 269) on the suitability of British limestones in various locations of a building (such as copings, parapets, strings, cornices, pavings and plain walling).

Sandstones are generally resistant to frost. Their general weather resistance is affected by their chemical composition. For example, calcareous sandstones in which sandstone grains are cemented by calcium carbonate are susceptible to attack by acidic rainwater. There are tests to verify the resistance.

Exclusion of moisture

One of the primary functions of external walls is the prevention of moisture ingress. The source of water includes rain and ground water. The following factors affect the resistance of a wall to wind-driven rain:

- presence of applied external finish
- quality of workmanship
- type of masonry unit
- mortar composition

- joint finish and profile
- thickness of the leaf
- presence of a cavity
- airspace within the cavity
- presence of cavity insulation.

BS 5628: Pt 3 provides guidance for the assessment of resistance to rain penetration through solid walls for different exposures. It also illustrates the effect of various factors affecting rain penetration of cavity walls.

Damp proof courses (dpcs) and cavity trays are provided to act as barriers to the passage of water from the exterior to the interior or from the ground to the structure or from one part of the structure to another. The passage may be horizontal, upward or downward. Careful architectural detailing is required for successful exclusion of water. In cavity walls, it is generally assumed that the rainwater will penetrate the outer leaf and run down the inside of the outer leaf. Thus there is a clear danger of the water being conducted across the cavity to the interior of the building when the cavity is bridged, e.g. cavity fill, lintels, structural beams, floor slabs or pipes. Watertight cavity trays are provided above all these bridges to avoid this problem. Weep holes in the outer leaf are incorporated at the cavity trays to allow water to escape. Wall ties inevitably bridge the cavity; here the safeguard is the twist in the tie, to act as a drip. BS 5628: Pt 3, again provides a number of illustrations of detailing dpcs and cavity trays.

Durability of masonry

Durability of masonry is mainly affected by two factors: frost action and sulphate attack.

Frost damages both masonry units and mortar, depending on their susceptibility to such damage on freezing in the saturated condition. The factors that affect the susceptibility of the masonry to damage are exposure to weather or other sources of water and the adequacy of measures taken to prevent the masonry becoming saturated.

Structures where masonry is likely to become and remain saturated include free-standing walls, parapets, chimney stacks, below the damp proof course at or near ground level, foundations and all substructure works.

It should be noted that neither strength nor water absorption of fired-clay masonry units are reliable guides for assessing the resistance to freezing. Past experience assisted by freezing tests will be a better guide. In calcium silicate bricks, strength and durability are related, and freezing and thawing has little effect on these bricks. Similarly concrete masonry units also possess excellent resistance.

When masonry units remain wet, expansion and deterioration of mortar can occur as a result of chemical reaction between soluble sulphates and constituents of cement in the mortar. The sulphates may be derived from ground waters or the ground, flue gases in chimneys or bricks containing sulphate-bearing clays. In these situations, it will be advisable to use sulphate-resisting cement mortars.

BS 5628: Pt 3 provides excellent guidance on the choice of masonry units and mortars for particular situations, based on durability considerations, and reference should be made to it.

Movement in masonry

Small dimensional changes in the finished construction are unavoidable. The design should therefor recognise this and allow for it. The main causes include:

- temperature changes
- changes in moisture content
- adsorption of water vapour (chemical bonding of water vapour molecules to masonry molecules)
- chemical actions
- deflection of supporting structures
- ground movement (differential settlements).

Generally movements caused by temperature and moisture changes are reversible. In all structures restraints are generally present that will prevent the free movement of masonry. As a result tensile stresses, and hence cracking, could develop.

Precautions should be taken to accommodate the movements and their effects. Fuller details may be found in publications such as BS 5628: Pt 3. One or two features are illustrated below:

- When the deflection of the supporting structure is the cause of the movement such deformation is limited to levels that will not cause cracking – normally (span/250) to (span/300). Vertical joints in the wall may also be necessary.
- When non load-bearing walls are built into concrete frames, it should be recognised that the fired clay bricks will tend to expand (due to adsorption) and that this will be opposed by the elastic deformation of the structure (including creep) and the shrinkage of the concrete. Horizontal movement joints will usually be necessary to prevent load transfer to the masonry.
- When movement joints are provided, they should be judiciously positioned and detailed. The width of the joint and any sealant used should be able to accommodate the expected movement.
- The provision of joints weakens the structure and also increases the risk of water penetration in external walls.
- The coefficient of thermal expansion of calcium silicate masonry units is substantially higher than that of fired-clay units. The shrinkage of autoclaved aerated concrete units is high compared to that of other concrete units or

calcium silicate bricks. Thus the spacing of movement joints in masonry using calcium silicate units or autoclaved aerated units will be smaller compared to those in fired-clay masonry.

• Bed joint reinforcement can be used to increase the resistance to cracking and hence the spacing of the joints.

Fire performance of masonry

Masonry possesses excellent fire resistance. After all, the process of manufacture of the units involves the heating of the materials to high temperatures. Units 100 mm thick will suffice for 2-hour fire rating, and 200 mm for 4-hour fire resistance periods. More precise details can be found in BS 5628: Pt 3.

One of the main problems with masonry in elevated temperatures, is that the unequal thermal expansion on the two faces causes thermal bowing. Perforated and hollow units are more sensitive to thermal shock than are solid units. Cracking of the webs in these units could lead to separation of the two leaves. All types of units give better performance when they are plastered.

In fire conditions, masonry may also suffer damage caused by the deformation of elements supported on it, e.g. expansion of floor slabs or unprotected steel structures can impose forces on walls. This can happen even in areas remote from the fire zone.

Structural timber

Introduction

Timber is a versatile material, used widely in structural and non-structural applications. It is also an old form of construction. In modern buildings it is widely used for floors and roofs. Completely timber-framed buildings are not yet common, but serious attempts are being made to promote them. Attractive domes, arches and portals using timber have been erected all around the world. Different forms of timber construction include solid timbers, glued and laminated members, plywood, chipboard and other particle boards.

In the UK, timber design is carried out using permissible stresses incorporating a global safety factor. Eurocode 5, dealing with the design of timber structures, uses limit state design for timber, in common with all Eurocodes. All timber codes recognise the peculiarities of the behaviour of timber and allow for them. Some of the major aspects are noted below.

Moisture content

The strength properties of timber increase or decrease as its moisture content is reduced or increased. Wood is less prone to decay if its moisture content is below 25 per cent and is generally considered immune below 20 per cent. Because of the effect of moisture content on strength, Codes of Practice prescribe different stresses for different exposures. In the UK only two categories are recognised in the code of practice BS 5268: Part 2. These are *dry exposure* and *wet exposure*. The former covers conditions of air temperature and humidity that would result in solid timber attaining an equilibrium moisture content not exceeding 18 per cent for any significant length of time. Most covered buildings and internal uses will come under this category. All other service conditions, where the equilibrium moisture content is likely to be in excess of 18 per cent for significant periods, come under the wet exposure category. Reduction factors for the permissible stresses are usually prescribed for this category.

Stress grading

As timber is an organic material, it is subject to wide variability because of genetic and species effects. If a single stress were to be used for each property of each species, such relatively low values would need to be specified as to make it uneconomic. Therefore a number of stress grades have been adopted. The grading method may be either visual or mechanical. In the former, basic properties for each species derived for straight-grained specimens, free from knots and fissures, are reduced by factors that account for strength-reducing effects of growth characteristics. In mechanical stress-grading, each piece to be graded is loaded centrally and the deflection is measured. The strength of the material is then calculated.

Duration of loading

It is known that timber structures can sustain much greater load for just a few minutes than for a longer period. This is recognised in Codes of Practice by multipliers for various periods of duration of loading. In the UK, four load-duration classes are used: long term (permanent loads), medium term (permanent + snow or other temporary imposed loads), short term (permanent + imposed + wind and/or snow) and very short term ((permanent + imposed + wind in 3–5 second gusts).

Direction of loading

Timber is idealised as an orthotropic material. Generally only two directions are considered: parallel to the grain and perpendicular to grain. The strength properties of the timber and the fasteners vary according to these directions and the variation has been found to follow the Hankinson relationship:

$$N = \frac{PQ}{\left(P \sin^2 \theta + Q \cos^2 \theta\right)}$$

where N is the stress at an angle θ to the grain, P the stress parallel to the grain and Q the stress perpendicular to the grain.

Load sharing

As the permissible stresses are derived statistically for single members, one could expect higher grade stresses when a load is supported by a group of members (the so-called load sharing system). Codes of Practice indicate the multipliers for stresses and modulus of elasticity.

Materials used in structural timber

Softwood and hardwood

In timber engineering, the timber species are broadly classified as: *softwoods* and *hardwoods*. The former refers to timber from conifers and the latter from deciduous trees. Some softwoods, e.g. Douglas fir, can be quite hard, and some hardwoods, e.g. balsa, can in fact be soft. The majority of routine structural applications use softwoods

Although the UK grows some softwoods, most is imported from countries such as Sweden, Finland, Canada and Norway. European whitewood and European redwood are two of the commonly-used softwoods.

The characteristic bending strength of softwoods is 14–40 N/mm^2. The 5 percentile values of the modulus of elasticity is in the range 4700–9400 N/mm^2. The mean density range is 350–500 kg/m^3.

For some applications, hardwoods are preferred on account of their greater strength and stiffness, higher density and fire resistance and their availability in longer lengths and sections. The increased costs should not be overlooked. The most well known hardwood is oak. The characteristic bending strength is in the range 30–70 N/mm^2 and the 5 percentile values of the modulus of elasticity are in the range 8000–16,800/mm^2. The mean density varies across 640–1080 kg/m^3.

Specifying solid timbers

In specifying timber, a number of points should be borne in mind. Although timber can be reduced to any size within reason, BS 4471 (Softwoods) and BS 5450 (Hardwoods) indicate the preferred industry norms.

The cross-sectional dimensions required in design should be stated on the drawings. For normal structural applications, the tolerances of timber as originally sawn from the log will be adequate. In some applications, all the surfaces of the members may need planing, and the planing tolerances should be taken into account in design.

The normal maximum length of softwood supplied is around 5.0 m. For longer members the length will need specifying. It is also advisable to discuss any non-standard requirements with recognised suppliers before proceeding with the design. It is essential to consult suppliers before specifying hardwoods.

Specification for strength can be either by species and the required grading, e.g. European redwood, GS (General Structural), or by a strength class, e.g. SC3. In the latter method the species of wood that will be supplied to the site will be the choice of the supplier. The above methods are in accordance with the current British practice. European standards are now being adopted and they use a slightly different method. The designation of the strength classes refers to the characteristic bending strength of the timber.

Glued and laminated timber (glulam)

Complicated shapes and curved structures are formed by gluing thin laminates of wood together. The maximum thickness of the laminates is usually limited to 50 mm. Members may be laminated either horizontally or vertically. Three grades are recognised in BS 4978 for horizontally laminated members. BS 5268, the code for the structural use of timber, gives stresses for each grade as a multiple of the properties of special structural (SS) grade timber of the same species. Where the member is built up using vertical laminations, the grade stresses appropriate to the species should be used. The special glulam grades do not apply to vertically laminated members.

The type of glue used should take into consideration the effects of the exposure, such as dry internal, humid internal (swimming pool) and external.

Plywood

Plywood is a form of glued-laminated construction formed by laying thin veneers with the direction of the grain in alternate layers rotated through 90°. An odd number of layers (*plies*) is used. This equalises the strength properties in two directions and also gives greater dimensional stability. In design there are two distinctly different approaches. The first is the *full area* method and the other is the *parallel ply* method. BS 5268: Pt 2 is written on the full area method.

Mechanical fasteners

Timber construction, like structural steel, mainly comprises discreet members that need connecting together to form a whole structure. While in some instances the fastener is simply a locating device, in many circumstances considerable design ingenuity is required. In the design of connections, the effects of end distances, edge distances and spacing between fasteners should be considered. It should be ensured that the mode of failure is that assumed in the member design and not a local premature failure at the connection. Often these considerations could govern the size of the member rather than the basic stress in the member.

The common types of fasteners used are nails, screws and bolts. The capacity of plain round wire nails can be enhanced by the use of *improved nails*. These include square twisting and square grooving, ring shanking and helical threading. The shear capacity of bolts can be enhanced by the use of timber connectors, of which there are three types: tooth-plate, split-ring and shear plate. In trusses, punched metal plates are commonly used. Thin metal framing anchors with pre-punched holes for nailing are extremely useful devices. Joist hangers are used extensively in timber engineering. It should be realised that, for stability, joist hangers rely on frictional forces developed by vertical loads. Therefore, it is inadvisable to use them at roof level. Also joist hangers apply eccentric loads on the supporting members and this should be allowed for in design.

The load capacities of nails, screws, bolts and connectors are given in Codes of Practice. Those of proprietary devices should normally be obtained from the relevant technical brochures. The design value is obtained by multiplying a basic value by factors to allow for considerations such as load duration, moisture content, length of joint and, in the case of connectors, for edge and end distances and spacing.

Most fasteners can be obtained with corrosion protection. Hot-dipped galvanised protection is the most common. Under some circumstances, metal fasteners may become corroded through contact with treated timber. The manufacturer of the chemicals used in the treatment should be consulted for any special precautions. Also some of the hardwoods, e.g. oak, are acidic and generally only non-ferrous fixings should be used for these.

Durability of timber

It is easy to realise why wood decays when it is appreciated that it is formed as part of a living plant and that it can provide nourishment for other living things, mainly certain fungi and insects. In drawing nourishment, these parasites destroy the wood. Fungal decay is initiated in wet conditions. The so-called 'dry rot' also requires wet conditions for its existence. In the final stage of dry rot attack, the wood is sufficiently dry and brittle to justify the name. Different types of fungi are at work in *dry rot* and *wet rot*. When it is remembered that timber is prone to fungal decay only when its moisture content is in excess of about 22 per cent, there should be very little need for special precautions with internal timbers in new construction. Careful detailing will be required to keep the timbers dry.

The four most common forms of insect attack encountered in the UK in softwoods are furniture beetles (woodworms), ambrosia beetles, wood wasps and house longhorn beetles.

Most structural timbers are protected by a preservative treatment, of which there are two principal types: water-borne salt types (solutions of copper, chromium or arsenic), which are also effective against the insects encountered in the UK, or organic solvent types, such as organic fungicides in an organic solvent. An insecticide could be added where necessary. The preservatives are applied by pressure or by vacuum and pressure.

Table 6.2 Rate of charring in Eurocode ENV 1995–1–2

		Density	Rate
(a) Softwood	Solid timber	≥ 290 kg/m^3	0.8 mm/min
	Glued and laminated timber	≥ 290 kg/m^3	0.7 mm/min
(b) Softwood	Solid or glued and laminated timber	≥ 450 kg/m^3	0.5 mm/min
(c) Hardwood	Solid or glued and laminated timber	≥ 290 kg/m^3	0.7 mm/min

Termites are a huge hazard in tropical areas of the world. However, unintentional importation and subsequent establishment of subterranean termites has been found in a localised area in Devon. As a result DETR have produced two BRE Digests 443: Parts 1 and 2, which contain guidance on detection, diagnosis and control of subterranean termites in the UK. For general advice regarding termite protection specialists should be consulted.

Fire performance of timber

Timber is a combustible material. It will survive only localised fires, where rapid intervention by the Fire Brigade has taken place or where the timber sections are massive. Timber 'browns' at around 120°C, 'blackens' at around 200°C and evolves combustible vapours at about 300°C. There will be no strength left in the charred layer. Below the charred layer it may be assumed that there is no significant loss of strength. The rate of charring is reasonably well established and is inversely related to density. Therefore for a given fire resistance period, the depth of charring can be estimated and discounted. Normal grade stresses can then be applied to the residual section. The rate of charring for various timbers given in Eurocode ENV 1995–1–2 (the part of Eurocode 5 dealing with fire) is listed in Table 6.2.

Metal connectors are vulnerable in fire conditions. They will require protection. Alternatively, the metal connections could be located below the line of the expected charring by suitable counter-sinking.

Impregnation treatments and surface coatings are available to retard the spread of flames over the surface of timber and wood-based sheet materials.

Further reading

Blake, L. S. (ed.) (1989) *Civil Engineers' Reference Book*, 4th edition, London: Butterworths.

BRE Digests on Concrete, particularly 237 *Materials for Concrete*, 244 *Concrete Mixes*, 250 *Concrete in Sulphate-bearing Soils and Ground Waters*, 258 *Alkali Aggregate Reactions in Concrete*, 263 *The Durability of Steel in Concrete: Part 1*, 330 *How to Avoid Alkali Silicon Reaction*, and 363 *Sulphate and Acid Resistance of Concrete in the Ground*.

BRE Digests 269 *Building Stones*, 380 *Damp-proof Courses* and 362 *Building Mortars*.

BS 5268: Parts 2 and 4, *Structural Use and Fire Resistance of Timber.*

BS 5328: Parts 1 to 4, *Guide to Specifying Concrete and Procedures for Producing, Transporting, Testing and Assessing Compliance of Concretes.*

BS 5493 and BS ENISO 12944, *Corrosion Protection of Steel Structures.*

BS 5628: Part 3, *Materials and Components, Design and Workmanship.*

BS 5950: Part 2, *Specification for Materials, Fabrication and Erection.*

BS 5950: Part 8, *Code of Practice for Fire Protection of Steelwork.*

Dowling, P. J., Knowles, P. and Owens, G. W. (1988) *Structural Steel Design,* Butterworths, London

ENV 206: *Concrete: Performance, Production, Placing and Compliance Criteria.*

Neville, A. M. (1981) *Properties of Concrete,* 3rd edition, Pitman.

Steel Construction Institute Publications, particularly 80 *Fire Resistant Design of Steel Structures* (A handbook to BS 5950: Part 8), 123 *Concise Guide to the Structural Design of Stainless Steel,* 119 *Design of Stainless Steel Fixings and Ancillary Components,* and 179 *Architect's Guide to Stainless Steel.*

The Prevention of Corrosion in Structural Steelwork, British Steel.

TRADA Publications, particularly 1–6 *Glued Laminated Timbers,* 1–17 *Hardwoods,* 2/3–10 *Timber Properties and Uses,* 2/3–16 *Preservation Treatment for Timber – A Guide to Specification,* 2/3–28 *Introducing Wood,* 2/3–35 *Adhesives for Wood* and 2/3–51 *Timber Engineering Hardware.*

Chapter 7

Structural safety, limit state design and codes of practice

Introduction: what do we mean by safety?

There can be no doubt that we should aim to design safe structures, but just what we mean by safety and how we attempt to ensure it is not so straightforward. Safety is one of those concepts that is difficult to define. Dictionaries give many shades of meaning but, for the purposes of structural design, we require a precise, mathematically acceptable definition. We shall work towards such a definition by discussing the problems relating to safety and how the treatment of safety has developed over the years.

We shall start with a simple illustration in order to obtain some idea of the basic problem faced by anyone who has to design or build a structure. Suppose that there is a stream that people need to cross. The stream is poisonous, or full of crocodiles, so falling in is fatal. It is decided to bridge the stream by laying a baulk of timber across. The average weight of a person is known reasonably well and we also have information on the average strength of the timber. Taking this average strength, the size of the baulk of timber required to support the average person can be calculated. If this were done and the bridge built, would you dare to cross it?

The answer to this question should be a resounding *no*! Half the people will weigh more than the average weight while half the possible baulks of timber will be below average strength. The first person to cross the bridge therefore has a 25 per cent chance of breaking the bridge and perishing in the deadly stream. Clearly the bridge must be made considerably stronger before it could be considered to be safe. It might be decided that, since absolute safety (whatever this may mean) is required, an enormously thick baulk of timber should be used. Unfortunately, this would cost more than the town council can afford. Somehow, a size of timber has to be chosen that will give an adequately safe bridge while not costing more than is in the budget. Striking a balance between strength and economy is the fundamental issue that has to be addressed when considering safety or, more generally, the quality of construction.

Any commercial organisation aims to maximise the profit it makes. If the firm's business is commercial construction then the pressure on those designing structures to minimise costs will be very great. An easy way to minimise costs is simply to reduce the amount of material used and hence reduce safety. However, firms also have a legal and moral duty not to put people at risk. Another example of the necessity to balance economy against safety or quality is the use of government resources to develop infrastructure (roads, hospitals, schools, etc). Tax revenue is limited and the pressure to economise in the standards and thus stretch the available money further will always be very high. For example, is it appropriate to build one very high-quality hospital or will society be better served by the construction of two hospitals of lower quality? Striking the right balance between quality, safety and economy is of fundamental importance and lies at the core of all design activity.

To strike such a balance in a rational and consistent way it is essential to be clear about what is meant by safety and to have a definition that is quantifiable. If this is not done then we are trying to balance a highly quantified concept (economy, measured in money) against a vague, qualitative concept of safety. In such circumstances, the balance will always tend to tip in favour of the quantifiable; consider, for example, the balance achieved in UK towns over the last 50 years between the vague qualitative concepts of beauty or aesthetic appeal and the pressures for economy.

Sources of uncertainty

Uncertainty about loading

To try to grasp how safety might be tackled, let us return to our bridge example. Instead of taking the average weight of a person and the average strength of the timber, it would seem a pretty obvious alternative to look for the heaviest person and the weakest timber and, using the lowest strength of timber, design the baulk to carry the heaviest person. Assuming these two factors can be established, the resulting structure will be safe and will not fall down, whoever crosses it. Unfortunately, this turns out to be an oversimplification of real life. The problem may be understood by considering the loading on a bridge and attempting to establish the maximum possible load that could be applied to it. In this case a real bridge designed to take motor traffic will be considered.

Thinking about the loading, we may conclude that there will generally be a mix of cars and lorries crossing the bridge and that, assuming the traffic is moving, there will be a reasonable gap between vehicles. This probably defines the normal loading, but what if there is a traffic jam and the lanes in both directions are now full of vehicles standing more-or-less bumper to bumper? Clearly the load will be greater. It is now necessary to consider how heavily loaded the vehicles are. Generally, most cars carry just one person, with some carrying as many as five with a boot full of luggage too. On average, we might expect about half the lorries to be fully loaded and the rest unloaded, since a lorry takes a load somewhere, unloads it and then returns. It is possible, however, that all the cars and lorries in a traffic jam

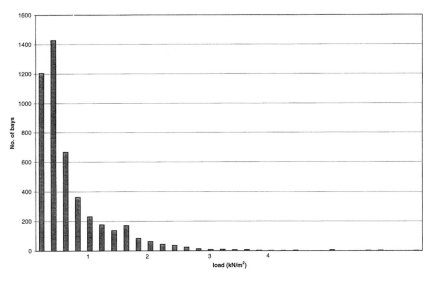

Figure 7.1 Frequency distribution of office floor loads on 14.5m² of floor.

Source: Mitchell, G. R. and Woodgate , R.W. (1971) *Floor Loadings in Office Buildings – The Results of a Survey,* Building Research Station Current Paper CP 3/71

could actually be fully loaded. It is also possible, though less likely, that there are no cars in the jam and that it consists only of fully-laden, bumper-to-bumper lorries. Is this, then, the maximum possible load? Unfortunately, it isn't. The police carry out checks on the weights of lorries on the road since there are legal limits on these and they always find that a significant proportion of the lorries they stop are overloaded. It could be possible, therefore, that all the lorries in the jam on the bridge are loaded to the maximum level measured by the police. Even this is not the end since we have not yet considered the footways. It is also possible that these are entirely packed with fat people carrying heavy rucksacks with more fat people sitting on the handrails. It will be seen that estimates of the maximum possible load that can be imposed on a bridge are limited only by your imagination. We have not yet, for example, considered the possibility that a herd of pigs has got onto the bridge and occupies all the space underneath the lorries or that a mercury tanker has crashed and filled all the drainage with mercury! The question: 'What is the maximum possible load that could be imposed on a bridge?' is simply unanswerable – it is not a useful question to ask. It should be observed, however, that, while it is possible to think of an infinite number of new ways in which more load might be put on the bridge, each succeeding scenario is more improbable than the last. What is actually required is to decide on the maximum loading that it is reasonable rather than the maximum load possible, and design for that. It may help the arguments if, at this stage, consideration is given to how loads can be studied to enable such a decision to be taken.

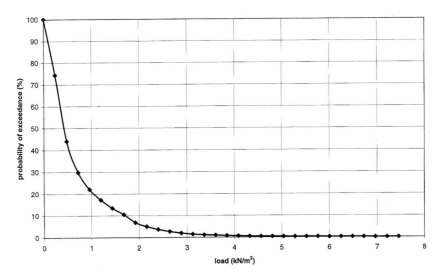

Figure 7.2 Data from Figure 7.1 replotted as a cumulative frequency diagram.

Clearly, it is necessary to carry out extensive studies of the loadings on structures of a particular type and then present the results in a way that will enable a view to be taken about the reasonableness of a given level of load. There are two basic approaches to this problem, depending on the nature of the loading. The first method is to survey the loadings on a large number of structures having a similar function; e.g. offices, retail premises or warehouses. These surveys can be used to produce histograms showing the relative frequency of occurrence of different levels of loading. As an example, Figure 7.1 shows the histogram of loads measured by Mitchell and Woodgate (1971) on office floors with areas of 14.5 m². This figure can be replotted as a *cumulative frequency* curve that shows the probability of the load exceeding a given value (Figure 7.2). This, however, is not an absolute maximum of the loads that might occur on this size of office floor. Had many more offices been surveyed, it is likely that floors with higher loads would have been found. To cope with this, a mathematical function may be commonly fitted to the upper end/*tail* of the distribution. A common approach is to plot the tail of the cumulative distribution on a logarithmic scale and then fit a straight line to it (Figure 7.3). This will permit better estimates to be made of the probabilities of occurrence of the higher loads (Table 7.1).

If, for example, we now decide that any probability of a load occurring that is less than 1 in 1000 is insignificant, then it can be seen from Table 7.1 or Figure 7.3 that the load we should take for design will be 6.1 kN/m².

The alternative way of studying loading is commonly used for environmental loads such as those from wind or waves. The approach is to obtain a continuous record of the loading from instrumentation and from this to establish *the return*

Table 7.1 Loading probabilities for 15 m² office floor slabs

Load (kN/mm²)	Probability of exceedence
2	4.3%
4	0.67%
6	0.11%
8	0.016%
10	0.0039%

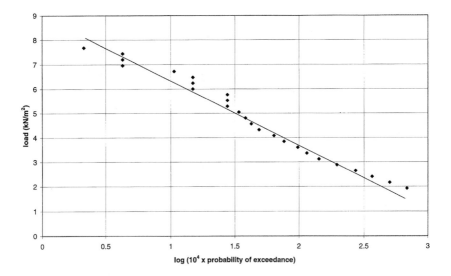

Figure 7.3 Tail of cumulative frequency diagram from Figure 7.2 plotted to give log probability of exceedence.

period for a given level of loading. The concept of return period requires a little explanation as, especially to the uninitiated, it is somewhat confusing.

The term suggests a periodic form of behaviour, but this is misleading. If, for example, a fifty-year return-period wind occurred yesterday, it does *not* mean that we will not get another such wind for fifty years. A simple example will explain the actual meaning. Supposing that a dice is thrown once a day. On each day there is a one in six chance of throwing a six or, put another way, there is a good chance that you will throw a six at least once during a six-day period. The return period for throwing a six is six days. The fact that you have thrown a six on one day obviously does not mean that you will not throw another six for the next five days; it is quite possible that you will throw another one tomorrow. A fifty year wind is thus one that has a good chance of occurring or being exceeded at least once during a fifty year period. There is no suggestion that it will occur only once or that a higher wind will not occur during this period. Wind speed is what is commonly measured, and wind speeds may be converted to wind forces by standard formulae.

Thus, in general, studies of loading cannot tell us what loads to use in design; they can only give information on the relative probability that various levels of loading will occur. The phrase 'in general' has not been used thoughtlessly here because there are some cases where the maximum possible loading can be defined with some precision. For example, a tank of water cannot be fuller than full and the density of water is known accurately. Such cases are, however, the exception rather than the rule.

Uncertainty about material strengths

So far, only loads have been considered. It is also necessary to look at the possibility of defining the weakest possible material.

If it were possible to test the strength of every bit of material that was included in a structure, one could guarantee the strength provided. Unfortunately, testing a material usually entails destroying it, so the actual material incorporated in the structure cannot be tested. The best alternative is to test some of the material and hope that the remainder, which is included in the structure, is similar. In statistical language, we sample the population and infer the properties of the total population from the properties of the sample. The population here is all the material which will be used in the structure. Since there is no possibility of knowing whether we have tested the weakest piece of material, we cannot know whether at a critical point in the structure there could be a weaker piece than we have tested. Hence, as for the loading, the question, 'What is the lowest possible strength for the material?' is unanswerable and the best that can be done is establish the probability that the strength is below a given value. There is an added risk with construction materials that the process of construction may cause the material to be weaker than that tested. For example, concrete has to be transported to where it is required, then placed, compacted, finished and cured. Any of these operations, if carried out imperfectly, can lead to the material being weaker than that sampled.

We have looked at two major aspects of design and construction: establishing the loading and establishing the material strength. It has been shown that there is a high degree of uncertainty about both, though it may be possible to say something about the probability that a given load or strength will occur. There are, unfortunately, yet more areas where uncertainty enters the process and some of these are discussed briefly below.

Uncertainty about structural dimensions

It is impossible to build a structure to exactly the sizes given on the drawings. There are recognised tolerances but measurements show that these are not infrequently exceeded. Thus, neither the spans of beams nor the member cross-sections will be exactly what was specified. In reinforced concrete, the position of the reinforcement will deviate somewhat from where it is shown on the drawings and all structures are likely to be somewhat out of plumb. All these factors will influence the strength, and need to be considered. Surveys have been carried out to establish the variability in the dimensions of structures and, for some situations, information is available of the probability that the dimensions will differ from the specified dimensions by specific amounts.

Uncertainty about accuracy of calculations

The behaviour of structures is highly complex and our understanding of this behaviour will always be incomplete. Furthermore, methods of calculation that most closely approach actual behaviour are likely to be very complex and time-consuming to apply. For this reason, simplifications and idealisations of the structure are always necessary in order to make the job of design possible and not unacceptably costly. As a simple example, when a structure is analysed to establish the distribution of forces and moments within the structure, it is normal to assume that all the members have stiffness but no dimensions other than length, i.e. the members are assumed to be concentrated on their centrelines. Furthermore, probably the materials will be assumed to be elastic. None of these assumptions is exactly true for any material or structure currently used and they will introduce errors into the results obtained. A more rigorous analysis could be carried out using methods such as three-dimensional finite element analysis. Such methods can also use the true (or at least more realistic) stress–strain curves for the materials. This approach, though possible with modern computers for relatively simple structures, would be prohibitively expensive. Furthermore, errors in the idealisations mentioned above are not usually too great and so the more accurate analysis is only very rarely justified.

Nevertheless, the idealisations introduce a further level of uncertainty into the design process. Incompleteness in knowledge is also a significant cause of uncertainty. The agreement between the prediction formula and the actual test strengths is often fairly good but is definitely never perfect. More important, while the performance of a formula may have been tested for common shapes of section, it is not known whether it would be as good for other sections that have not been covered by the research. Shear of reinforced concrete members without shear reinforcement, for example, is particularly susceptible to this type of uncertainty because no generally accepted theory exists to explain behaviour and formulae are mostly based on tests rather than any deep understanding of the phenomenon.

Effect of time

All materials change with time and the effects of the environment to which the structure is exposed. Concrete, for example, generally increases in strength with time but, in some circumstances, can deteriorate. Steel rusts unless protected by paint; wood may rot, if damp, and other materials can become brittle with time.

Means of providing safety taking account of uncertainty

Early developments

There is uncertainty about the loading, uncertainty about the strengths of the materials that will be used, uncertainty about the quality of construction, uncertainty about our ability to predict behaviour. Despite all these, we still have to design and construct safe and economical structures. How, in practice, can we rationally allow for these uncertainties? The pressures for maximum economy have increased over the years, and so ideas on how to provide safety have tended to become more sophisticated. The various approaches may therefore conveniently be explained by setting out the history of their development from the most primitive to the current situation with, possibly, a look to the future.

Until the beginning of the nineteenth century, very little design was carried out using any form of calculation. Means did not exist for the measurement of the strengths of construction materials and nor had the theoretical knowledge of the behaviour of structures developed to a level where such information could have been used to predict the carrying capacity of structures, had it been available. Safety was assured by the application of experience. A builder, carpenter or mason had spent his life gaining experience about what forms of structure, what materials and what dimensions would work in particular circumstances. This knowledge was passed down from master to pupil by the apprenticeship system and had been built up over centuries (or even millennia). In general the system was reliable and produced structures that served their purpose safely; some of them are still doing so. A weakness in the reliance on experience is that, while designs that were obviously unsafe would not be built, there was no means of knowing whether the strength provided was excessive and therefore whether a more economical solution would have been possible. A more critical requirement is that the system works only where conditions are roughly static; that is to say where the economics of construction do not change and where the materials used and types of structure required do not change greatly. These conditions held until the Industrial Revolution. The first half of the nineteenth century saw radical changes in all these areas. In 1790 Abraham Darby built the first cast iron bridge at Ironbridge. He had absolutely no experience of how to build cast iron bridges since his was the first. Inspection of the bridge, which still stands, shows that he copied many of the details from timber technology, though this is hardly likely to have been really relevant. Materials developed rapidly over the next half century with wrought iron and, eventually, steel entering the construction field. Towards the end of the century, reinforced concrete added another, completely new material to the construction industry's resources. At the same time as the materials available to the construction industry were changing, the types of structure required were also changing radically. The development of the railways, and their requirements for very limited gradients, led to a requirement for huge number of bridges with spans well beyond those for which experience could cater. The loads that they were asked to carry were also beyond current experience. In other areas, the development of factories and mills led to requirements for new types of buildings, as did the development of other infrastructure areas such as ports. These challenges were quite beyond the experience of the designers and builders and new methods had to be found. These were the pressures that eventually led to the development of structural engineering as we know it today.

Early attempts to develop safe designs tended to involve testing of the new structures. Thus Brunel, when he built his railway bridge over the Wye at Chepstow, erected one of the girders on temporary supports and tested it under a distributed load of 770 tons prior to installation. Robert Stephenson used model testing extensively in the design of the Britannia Bridge. Their approach to safety was thus to ensure that their structures were able to support a load substantially larger than that which was actually expected. This can be expressed symbolically as follows:

$$R \geq \gamma S$$

where R is the calculated resistance of the structure (strength), S is the assessed design loading and γ is a safety factor.

The purpose of the safety factor is to allow a margin to take account of all the uncertainties in the assessment of the loads and the strength. The safety factor was arbitrarily chosen, commonly taking a value of around 3. This is a very convenient formulation of the safety problem where either the strength has been found by experiment or the structure has been tested up to some multiple of the design load.

The permissible stress approach

During the nineteenth century, great strides were made in the development of the theory of structures. The works of such famous people as Navier, Rankine, Clerk Maxwell and Castigliano all date from this period. What was developed was the ability to carry out the elastic analysis of structures and calculate the stresses in elements. The elastic theory does not directly lead to a prediction of strength, but of the distribution and magnitude of stresses. To make an assessment of strength it is necessary to establish the maximum stress that the materials can support. This led to the formulation of the safety problem in a rather different way. The ultimate stress that a material could support was divided by a safety factor to give a 'permissible stress'. Calculations were then carried out to ensure that this permissible stress was not exceeded. This approach may be expressed symbolically as:

$$\sigma_R/\gamma \geq \sigma_s$$

where σ_R is the maximum stress sustainable by the material, σ_s is the maximum stress calculated to occur under the design load, and γ is a safety factor.

If the permissible stress is defined as σ_p where $\sigma_p = \sigma_R/\gamma$, then this reduces to:

$$\sigma_p \geq \sigma_s$$

This is generally known as the *permissible stress* formulation of the safety problem. Again, the value of the safety factor was chosen arbitrarily with no attempt to establish what it should be as a function of the uncertainties involved. The values for the design loadings from which the stresses were calculated have generally been taken as working loads: that is, loads that have a serious chance of being encountered during the life of the structure. An advantage of adopting this level of loading is that the loads look realistic to the designer and the client, for example, this may be the load that the client actually expects to put in his warehouse. Also, under these loads, the structure probably does behave roughly elastically so the calculation of stresses and loads are probably relatively realistic.

Developments to the permissible stress approach

It is at about this point in time that reinforced concrete arrived as a significant structural material. Its development introduced a new complication into the safety problem. Reinforced concrete is a composite material made of steel and concrete acting together. These two materials are very different in the way they are produced and therefore in the uncertainties that might be expected about their strengths. Reinforcing bars are produced by sophisticated industrial processes under factory control. On site they are merely bent and fixed into position, neither process being expected to make major alterations to the basic material properties. Concrete, on the other hand, is largely made from gravel and sand, which are natural materials. In the early days of reinforced concrete, the concrete was mixed on site, probably by men with shovels. It was then placed into the moulds and tamped to compact it. The control that could be applied to the batching, mixing and placing process was limited and the result could be decidedly variable. It was rapidly realised that the permissible stress approach could be adapted very easily to take account of different levels of uncertainty in different materials simply by using different safety factors for the different materials. The permissible stress format for multiple materials could now be expressed as follows:

$$\sigma_{R1}/\gamma_1 \geq \sigma_{s1}$$
$$\sigma_{R2}/\gamma_2 \geq \sigma_{s2}$$
$$\sigma_{R3}/\gamma_3 \geq \sigma_{s3}$$

and so on, where σ_{R1}, σ_{R2}, and σ_{R3} are respectively the ultimate stresses sustainable by materials 1, 2 and 3; γ_1, γ_2 and γ_3 are respectively the safety factors for materials 1, 2 and 3, and σ_{s1}, σ_{s2} and σ_{s3} are respectively the maximum stresses calculated in materials 1, 2 and 3 under the design loading.

This approach is clearly quite sophisticated in its ability to handle materials of different strengths and with different related uncertainties. The formulator of such a method now has a more complex task, since some judgement needs to be made about the relative uncertainties, whereas in previous methods it was only necessary to decide on some global figure that could be expected to cover all sources of uncertainty.

The partial safety factor method

A weakness with the above approach is that, while it recognises that different materials have different uncertainties associated with them, it does not give any recognition to the different uncertainties associated with different types of loading. A second weakness in the way the permissible stress formulation was generally used is that there is not necessarily a direct relationship between stress calculated elastically and failure load. The safety factors are thus notional and do not necessarily relate closely to the true safety of the structure. During the middle years of the twentieth century, much research was done on the prediction of the strength of members and, in considering safety, it seemed more logical to many people if the safety problem was formulated in terms of ultimate strength rather than service conditions. In a number of design codes, e.g. CP114 (1969), ultimate strength equations were introduced but were disguised to look like service stress equations. The reason for this subterfuge was that designers were familiar with the old service stress values and sudden changes in stress or load levels could lead to misunderstandings and mistakes. To resolve these perceived weaknesses, a new formulation of the safety problem was developed. It is believed to have first appeared in Russia in the 1930s, but was first developed into its present form in the Recommendations of the Comité Européen du Beton of 1964. This was the *partial safety factor* format.

The initial step in the formulation of the partial safety factor method is the development of a rigorous definition of the loadings and strengths of the materials. The discussion so far has carefully avoided this. This principal representative value of the load is called the *characteristic value*. Eurocode 1 (1994) states that the characteristic load, '... is chosen so as to correspond to a prescribed probability of not being exceeded on the unfavourable side during a "reference period" taking into account the design working life of the structure ...'. Possibly a very low probability of being exceeded should be chosen, such as 1 in 10 000, or possibly one should select a load with a much higher probability of exceedence, such as the mean value. In fact, it seems most convenient to take as the definition of characteristic loading a level of loading that could realistically be expected to occur some time during the lifetime of the structure. The reason for this choice is that it is effectively the level of loading that has traditionally been adopted and is therefore a level of loading that is familiar to designers. Some documents have suggested that the 'prescribed probability' should be five per cent, though this will not normally be found in any codes of practice.

A similar approach may be adopted for defining the strength of construction materials, and Eurocode 1 defines the characteristic property of a material as: 'The value of the material property having a prescribed probability of not being attained in a hypothetical unlimited test series. It generally corresponds to a specified fractile of the assumed statistical distribution of the particular property of the material.' In this instance, codes and standards are usually clear that a five per cent fractile is intended. For example, BS8110 gives the definition as: 'that value of the cube strength of concrete, the yield or proof strength of reinforcement or the ultimate strength of prestressing tendon below which five per cent of all possible test results would be expected to fall.'

The reference in these definitions to a 'hypothetical unlimited test series' or 'all possible test results' is required to take account of the fact that the strengths can only be sampled and the properties of the total amount of material inferred from this sample. The phrases quoted are supposed to indicate that it is the inferred properties of the total material that are required and not the properties of the sample. There are standard statistical methods of correcting the properties obtained from a sample to give an estimate of the properties of the total population. Provided a reasonable number of samples are tested, the differences are not great.

Having defined characteristic loads and strengths, these may now be used in the definition of design strengths and loads that will have much lower probabilities of occurrence. The design strength of a material is given by the relation:

$$f_d = f_k/\gamma_m$$

where f_d is the design strength of the material considered, f_k is the characteristic strength of the material considered and γ_m is the partial safety factor appropriate to the material considered.

The design strength of the section, member or structure may be calculated as a function of the design strengths of the materials. This may be written symbolically as:

$$R_d = f_n\{f_1//\gamma_{m1}, f_2//\gamma_{m2}, f_3//\gamma_{m3}, \text{etc.}\}$$

where R_d is the design strength, and $f_n\{\}$ indicates a function of the quantities in the brackets.

This is, so far not very different from the permissible stress approach except that ultimate strengths are being considered instead of stresses under service loads. The significant difference from the permissible stress formulation is in the way that loads are treated. Here each individual load is multiplied by a partial safety factor that reflects the uncertainty in its determination. Design loads are defined by the relation:

$$F_d = \gamma_f F_k$$

where F_d is the design load, F_k is the characteristic load and γ_f is a partial safety factor appropriate to the particular loading.

It is not necessarily useful to consider the total load on a structure to be the sum of the various design loads, as some loads may act in different directions from others (for example, some may be vertical while others, such as wind, may act horizontally), so it is usual to work in terms of the design effect of the loads on the structure S_d. Thus:

$$S_d = f_n\{\gamma_{f1} F_{k1}, \gamma_{f2} F_{k2}, \gamma_{f3} F_{k3}, \ldots\}$$

To clarify slightly further what is meant by the 'effect of the loads' or *load effects* as they are sometimes called, it must be seen that any load or combination of loads acting on a structure induces in that structure a set of internal forces and moments that exactly balance the applied loads. These internal forces are the load effects. They may be direct forces, moments, shears, torsions or any combination of these.

The design requirement for a safe structure may now be written as:

$$R_d \geq S_d$$

The notation used above is that used in the Eurocodes, as this appears to be the most consistent set currently in use. The notation in UK codes is slightly different but should be readily understood.

A problem with this formulation is that there are more uncertainties than just the uncertainty about the material strength in the structure and the loading. Formulations can be proposed where other factors, such as dimensions, also have characteristic values and associated partial safety factors, and those further factors add to the uncertainty about the accuracy of design equations. This, however, is generally considered to be unnecessarily complicated. As a result, the values chosen for γ_f and γ_m are assumed to include an allowance for these other factors. The basic definition of the partial factors is that γ_f takes account of all factors that influence the precision of the calculation of the action effects (i.e. the size and distribution of direct forces, moments, shears or torsions within the structure), while γ_m takes account of all the factors that influence the precision of the calculation of the resistance of the structure. If all sources of uncertainty are listed, the uncertainties included in the two factors are as follows.

γ_f covers:

- uncertainty about the magnitude of the loading
- uncertainty about the distribution of the loading
- uncertainty about the method of prediction of the action effects (structural analysis)
- uncertainty about the structural dimensions in so far as these affect the estimation of the action effects.

γ_m covers:

• uncertainty about the strengths of the materials as tested

• uncertainty about the differences between the material as tested and the material as incorporated in the structure

• uncertainty about the structural dimensions in so far as this affects the calculation of the resistance

• uncertainty about the accuracy of the methods of prediction of strength.

It will be seen that, in an attempt to provide a more rational treatment of what can now be seen to be a very complex problem, the formulation of the safety problem is becoming much more sophisticated and complex.

Probabilistic design

At about the same time as the partial safety factor method was being devised, thought was being directed towards a revolutionary new approach to the problem whose rigour would be limited only by the data available on the uncertainties being considered. This is the fully probabilistic method.

The first essential for a probabilistic design method is a definition of safety, and the definition used is that safety is related to the probability of failure of the structure. The lower the probability of failure, the higher is the safety. The starting point for the method is now to specify an acceptable probability of failure. Design will then be carried out to ensure that the probability of failure does not exceed this limiting value.

The concept of the approach is illustrated schematically in Figure 7.4. For each element in the design where there is uncertainty, information must be gathered and a frequency distribution established. A frequency distribution for imposed loads on office floors was discussed earlier and is shown in Figures 7.1 to 7.3. These frequency distributions are shown schematically at the top and bottom of Figure 7.4. Starting from the top, the loads and the dimensions and material stiffnesses are used in an analysis method to establish the frequency distribution of the load effects. This has to be modified to allow for the uncertainty introduced by the analysis method itself. From the bottom of the figure, the material properties and dimensions are used as input data to a method for calculating a frequency distribution for the structural resistance. This has to be modified to allow for the uncertainty inherent in the calculation method and any uncertainties introduced by the construction process. The problem now is to estimate the probability of the load being greater than the resistance. A convenient way of doing this is to rearrange the basic safety inequality $R \geq S$ to $R - S \geq 0$ and to produce a frequency diagram for the quantity $(R - S)$. Proof of the adequacy of the design now requires that the probability of $(R - S)$ being less than 0 does not exceed the specified value.

If the data exists for each of the variables, and if the frequency distributions can be approximated to one of the standard forms of frequency distribution used in

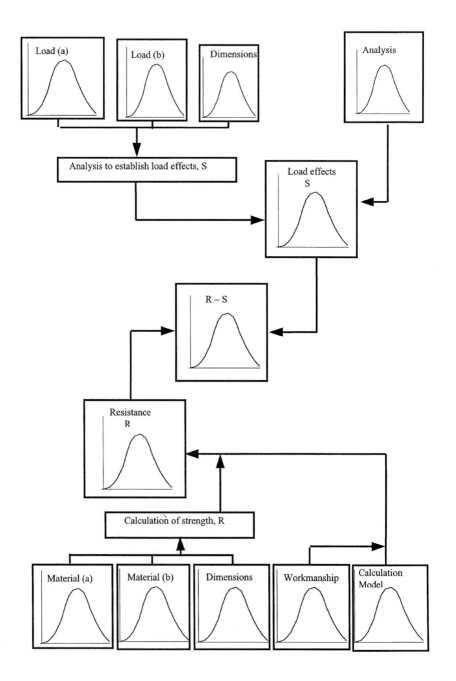

Figure 7.4 Concept of probabilistic design.

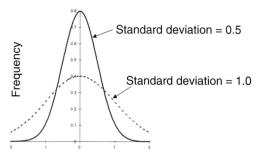

Figure 7.5 Variation in the shape of a frequency distribution as a function of the standard deviation.

statistics, then there are standard statistical procedures that will enable these calculations to be carried out. For example, a normal distribution is a symmetrical bell-shaped distribution given by the equation:

$$y = \frac{1}{\sigma\sqrt{2\pi}} e^{-\frac{(x-\bar{x})^2}{2\sigma^2}}$$

where y is frequency, x is the value of ordinate at which y is to be calculated, $\bar{x}$ is the mean value and σ is the *standard deviation* of distribution.

The standard deviation is a measure of the spread of the distribution. For any frequency distribution built up from n measurements, the standard deviation is given by:

$$\sigma = \sqrt{\frac{\sum x^2}{n} - \left(\frac{\sum x}{n}\right)^2}$$

Figure 7.5 illustrates how the shape of the distribution changes with change in standard deviation.

The variability of many common natural factors, e.g. the variation in height of the population, approximates to a normal distribution and it is often found to give a reasonable fit to data such as the variations in strength of materials.

If we assume that we have calculated the frequency distributions of R and S in Figure 7.1, and these can be assumed to be normal, then the distribution of (R – S) will also be normal with a mean value of $(R_{mean} - S_{mean})$ and a standard deviation of $\sqrt{(\sigma_R^2 + \sigma_S^2)}$. It is convenient to express the safety in terms of the *safety index* β, where:

$$\beta = (R_{mean} - S_{mean})/\sqrt{(\sigma_R^2 + \sigma_S^2)}$$

Table 7.2 Probability of failure as a function of the safety index, β, for a normal distribution

β	Probability of failure
1.64	0.05
2.0	0.023
3.0	0.003
4.0	0.0009
5.0	0.00000

Provided the distribution is normal, there is an exact relationship between β and the probability of failure $(R - S) < 0$. This is shown in Table 7.2.

A very simple example will show how a design approach might work. Assume that it is required to support a load that has an average value of 100 kN with a standard deviation of 40 kN by hanging it on a bar with a tensile strength of 400 N/mm^2 with a standard deviation of 40 N/mm^2. What cross-sectional area of bar will be needed if a safety index of 5 is to be achieved?

The strength of the hanger is given by the strength of the material multiplied by the cross-sectional area of the hanger, A. If the cross-sectional area is assumed to be subject to no uncertainty then we can substitute into the equation above for the safety index to give:

$$5 = (400A - 100\,000)/\sqrt{((40A)^2 + 40\,000^2)}$$

The cross-sectional area A is the only unknown, and the equation can be solved to give $A = 934$ mm^2.

It is possible to carry out similar calculations using other distributions, or for more complex relationships between variables. It is only intended here to suggest that the approach is possible, if the data exist.

This may be considered to be the ultimate safety approach: design may be carried out to achieve any specified probability of failure provided that the uncertainties in all the variables are known and can be characterised by frequency distributions. Unfortunately, the calculations involved are often very complex and, more critically, full data on the uncertainties are not available. We are therefore a long way from being able to implement such a system at present. Also, for reasons that will be discussed later, it can be argued that such a system can never properly deal with some of the uncertainties in real design situations. It may, therefore, never be worth trying to implement a fully probabilistic system in practice.

What can be done, and is being done to an increasing degree, is to use probabilistic calculations and statistical data to 'calibrate' partial safety factors. This is not an exact process but should enable a partial safety factor method to achieve a more uniform level of safety. Part 1 of the new Eurocode 1 (1991) covers the basis of design and includes a method for carrying out this calibration.

This completes what may appear to be a rather complex discussion of the methods for ensuring that designs attain an adequate level of safety. It has to be said that appearances are deceptive and the description has, in fact, simplified to a great degree some very complex issues. Nevertheless, sets of equations are confusing unless it is shown how they may be applied. The actual application will then be seen to be fairly straightforward in all cases except the fully probabilistic approach. To illustrate this, and to obtain further insights into the relative merits of the various approaches, a series of examples will now be presented.

Examples of use of safety formats

You are probably now somewhat bemused by all the different ways of handling the safety problem, and their relative merits. To try to help clarify the issues, this section gives examples of the various methods in use and illustrates their strengths and weaknesses. The approach adopted is to see how the various methods handle a series of generic problems. The problems to be studied are given below. For the purposes of this exercise, all variables will be assumed to have properties that are normally distributed and therefore can be uniquely defined by a mean value and a standard deviation. The basics of the probabilistic design method have been discussed in the foregoing section. Use of the method is often complex and requires knowledge of standard statistical formulae. Calculations using this method will therefore not be set out in full, but the results will be given and these will be used as a basis for judging the performance of the other methods. The base values given in Table 7.3 will be used throughout.

In use of the probabilistic approach, a safety index of 4.5 has been aimed for.

In the application of the partial safety factor method, the partial safety factors used are as given in Table 7.4.

For the global load factor method, a factor of 1.6 has been selected.

For the permissible stress method, partial factors of 1.7 for concrete and 1.4 for steel have been chosen.

Table 7.3 Base values for comparison

Variable	Mean value	Standard deviation	Characteristic value
Bad diameter	25 mm	1 mm	
Depth of cross-section	350 mm	5 mm	
Breadth of section	350 mm	5 mm	
Strength of steel	500 N/mm^2	40 N/mm^2	434 N/mm^2
Strength of concrete	40 N/mm^2	6 N/mm^2	30.2 N/mm^2
Dead load	$G_m = G_k$	$0.25G_k$	G_k
Imposed load	Q_m	$0.5Q_m$	$1.825Q_m$

Table 7.4 Partial safety factors used in comparisons

Variable	Partial factor
Concrete strength	1.3
Steel strength	1.1
Dead load	1.2
Imposed load	1.4

None of these values relates to any specific code of practice or set of design rules, but have been chosen to give a general consistency of safety level across the methods for Problem B.

Problem A Single load, single material

Consider the safe load capacity of a 25 mm diameter steel bar loaded in axial tension. The whole of the load is assumed to be imposed. The capacities by the various methods for this problem are as follows:

1 Global safety factor

$$\gamma Q_k = f_{yk} \, \pi\phi^2/4$$
hence $\quad 1.6Q_k = 434\pi 25^2/4$
hence $\quad Q_k = 133$ kN

2 Permissible stress (a) and (b)

$$Q_k = f_{yk} \, \pi\phi^2/4\gamma$$
hence $\quad Q_k = 434\pi 25^2/(4 \times 1.4)$
hence $\quad Q_k = 152$ kN

3 Partial factor

$$\gamma_q Q_k = f_{yk} \, \pi\phi^2/4\gamma_{ms}$$
hence $\quad 1.4Q_k = 434\pi 25^2/(4 \times 1.1)$
hence $\quad Q_k = 138$ kN

4 Probabilistic

$$Q_k = 60.5 \text{ kN}$$

Problem B   Two load types on two-material composite

This is the classic reinforced-concrete problem. Here, the load capacity of an axially loaded 350 mm square column will be considered where the dead load is equal to the imposed load. The ultimate strength of this column will be given by the equation:

$$N = f_y A_s + 0.6 f_{cu} A_c$$

1 Global factor

$\gamma(G_k + Q_k) = f_y A_s + 0.6 f_{cu} A_c$
hence $1.6(2 \times Q_k) = 434 \times 1964/1000 + 0.6 \times 30.2 \times 350 \times 350 / 1000$
hence $Q_k = 960$ kN

2 Permissible stress (a)

n = Where the elastic permissible stress method is used, a different equation is necessary since the stress in the concrete is always related to the stress in the steel by the equation:

$$f_c = f_s E_c / E_s$$

Traditionally, the modular ratio E_c/E_s is taken as 15, and this has been used here. The design service stress is now taken as the lesser of the two values given by the equations below:

$(G_k + Q_k) = f_{yk}/\gamma_s A_s + f_{yk}/15\gamma_s A_c$
or $(G_k + Q_k) = 15 f_{cu}/\gamma_c + f_{cu}/\gamma_c A_c$
hence $2Q_k = 434/1.4 \times 1964/1000 + 434/(1.4 \times 15) \times 350 \times 350/1000$
hence $Q_k = 1570$

or

$2Q_k = 30.2/1.6 \times 15 \times 1964/1000 + 30.2/1.6 \times 350 \times 350/1000$
hence $Q_k = 1434$
hence correct answer = 1434 kN

3 Permissible stress (b)

$(G_k + Q_k) = f_{yk}/\gamma_s A_s + 0.6 f_{cu}/\gamma_c A_c$
hence $2Q_k = (434 \times 1964/1.4 + 0.6 \times 30.2 \times 350 \times 350/1.6)/1000$
hence $Q_k = 998$ kN

4 Partial factor

$(G_k\gamma_g + Q_k\gamma_q) = f_{yk}/\gamma_s A_s + 0.6\, f_{cu}/\gamma_c A_c$
hence $Q_k(1.2 + 1.4) = (434 \times 1964/1.1 + 0.6 \times 30.2 \times 350 \times 350/1.3)/1000$
hence $Q_k = 954$ kN

5 Probabilistic

$Q_k = 910$ kN

Problem C   Simply supported beam with cantilever

This is the classical stability problem. The issue is not the strength of the members in the structure, but simply the balance of the loads. The problem is to find the load that can safely be put on the cantilever before the right-hand end of the beam lifts off its supports (Figure 7.6).

The beam supports a dead load over both spans and, in the worst situation, will be subjected to an imposed load on the cantilever only. By taking moments about the support at the root of the cantilever, it can be seen that the condition for failure by overturning is:

$$L_1(G + Q)/2 > L_2 G/2$$

Dividing both sides by $L_1 G$ gives:

$$1 + Q/G > L_2/L_1$$

In the following it will be assumed that $L_2/L_1 = 5$ and the objective is to find the ratio Q_k/G_k that the structure can withstand without failing.

1 Global safety factor.

Since both Q_k and G_k are effectively multiplied by the same safety factor, then the safety factor cancels and $Q_k/G_k = L_2/L1 - 1 = 5 - 1 = 4$

Figure 7.6 Problem C.

2 Permissible stress methods (a) and (b)

Since no safety factors are applied to the loads, both methods will give $Q_k/G_k = 4$, as above.

3 Partial safety factor method

In this case, the design equation becomes:

$$\gamma_q Q_k/\gamma_g G_k = L_2/L_1\gamma_{gmin}$$

γ_{gmin} is a figure less than 1, which reflects the minimum value that the dead load might take. A reasonable assumption here might be to take $1/\gamma_g$ and this is what has been done. It has to be said that, in many codes, this minimum value is taken as 1.0.

$$1.4Q_k/1.2G_k = 4 \times 0.8333$$
$$Q_k/G_k = 2.86$$

4 Probabilistic

There are problems in deciding what to do here, as we might expect there to be a degree of correlation between the dead load of the cantilever and that of the beam, and there is no means of guessing what this might be. Assuming full correlation, the probabilistic approach gives $Q_k/G_k = 2.24$.

Problem D   Axially loaded pinned ended strut

The capacity of an axially-loaded pinned-ended strut is given by the well-known *Euler equation* as:

$$P = \pi^2 EI/L^2$$

where P is the maximum supportable axial load, E is the modulus of elasticity of the material, I is the second moment of area of the cross-section of the strut, and L is the overall length of the strut.

It will be seen that the strength of such a strut is independent of the strength of the materials from which it is made. This example is somewhat academic since it is not generally possible to ensure pure axial loading and some initial eccentricity is usually included in practical design methods. This changes the nature of the equations. Nevertheless, it seems a valid test of the weaknesses and strengths of the various safety formats.

1 Global safety factor

Since the safety factor is applied to the loads, a safety factor of 1.6 is obtained.

2 Permissible stress methods (a) and (b)

Since the safety factor is applied to the material strengths, which do not appear in the prediction equation, no safety margin is obtained and the safety factor is 1.0.

3 Partial safety factor method

The partial factors on the material strengths provide no safety, since the strengths do not appear in the prediction equation. However, the partial factor on the loading will have an effect and, assuming that the imposed load is equal to the dead load, an overall safety factor of 1.3 will be obtained.

4 Probabilistic method

As the question has been posed, the statistics of the loading and the variability of the section dimensions, which will influence the value of I, will influence the calculated design load capacity. With the variables used, this will result in an overall factor of safety of 1.65.

Summary

Table 7.5 tries to summarise the performance of the various methods relative to the probabilistic approach. In the table, the capacity predicted by the various methods divided by the capacity predicted by the fully probabilistic method is tabulated for each method for each of the four problems. A value greater than 1.0 indicates an unsafe answer. An attempt was made to calibrate the partial factor method, the global safety factor method and permissible stress method (b) so that they gave approximately the same answer as the probabilistic method for Problem B. It will be seen that the rounding of the factors that were chosen has led to an underestimate of the answers to B. This has been allowed for in the table, so that Problem B gives an answer of 1.00 for these three methods.

Table 7.5 Summary of comparisons

	Problem A	Problem B	Problem C	Problem D
Global safety factor	2.08	1.00	1.70	1.00
Permissible stress (a)	2.37	1.50	1.70	1.57
Permissible stress (b)	2.26	1.00	1.70	1.57
Partial safety factors	2.15	1.00	1.22	1.21
Probabilistic method	1.00	1.00	1.00	1.00

The conclusion is clearly that none of the methods is perfect, but that the partial safety factor method is probably the best. All methods give poor predictions for Problem A. The reason for this is that a safety factor method calibrated to deal with a two-material problem does not generally work well for a single-material problem. Thus different calibration exercises would be needed for the development of a structural steel design method, which is basically a single-material problem, then those needed for reinforced concrete where the problem is a two-material problem.

Problem C is poorly tackled by both the global safety factor method and the permissible stress method. This poor performance on stability questions is one of the main reasons often quoted for the change to the partial factor method in UK codes in 1972.

Overall, the worst performance is for the permissible stress method based on elastic design.

Accidents and robustness

At 5.45 on the morning of 16 May 1968, Mrs Hodge, the tenant of a flat on the 18th floor of a block of flats in Canning Town (called Ronan Point), went into her kitchen to make herself a cup of tea. What she didn't know was that there was a faulty connector in the gas pipe to her cooker that had led to gas leakage during the night and that an explosive mixture of gas and air had built up. When she struck a match, an explosion occurred which led to the collapse of an entire corner of the block. Ronan Point was constructed of precast concrete wall panels and floors. The explosion was sufficient to blow out much of the kitchen wall, thus removing support from the floor above. This fell and the debris loading caused the floor of Mrs Hodge's flat to collapse onto the floor below. This resulted in an increasingly large falling mass of rubble that demolished each floor in turn down to ground level. The collapse of Ronan Point is the classic case of *progressive collapse* or the 'house of cards' type of failure. The collapse, together with two other failures about the same time (the roof of a school hall in Camden Town and a building that fell during construction at Aldershot), led to much debate and to the development of new design concepts. There are three issues relevant to this type of situation that need to be considered in relation to structural safety. These are: what forms of loading we should design for, accidents and 'acceptable modes of failure'.

Accidents are unforeseen and therefore we do not design specifically for the loads that they might impose on structures. When Mrs Hodge's kitchen was designed, no consideration was given to the pressures that it might be subjected to if the gas cooker blew up. It might be asked whether this possibility should have been foreseen and the structure designed to resist it. Gas explosions constitute a rather difficult case and we shall return to these later, but first let us consider accidents more generally. By definition, accidents lie outside the general operating conditions of a structure; they are unforeseen and the range of possible accidents is huge. For example, an aeroplane might fly into a structure. This can happen: a bomber flew into the Empire State Building in New York and a passenger jet flew into a block of flats in Amsterdam a few years ago. Should we design for this? The answer would seem to be that whether a particular type of accident should be designed for or not will depend on the balance between the cost of designing against a particular form of accident, the likelihood of it occurring and the possible consequences of such an accident. As an example, the possible consequences of collapse of the containment vessels in a nuclear power plant are almost unthinkably severe; such vessels are designed for earthquakes, while normal construction in this country is not. The possibility of aeroplane impact is also considered for nuclear installations. There is, in fact, a method for assisting in making decisions about what hazards to design for. This is *risk analysis*, which is a systematic approach to establishing all possible accident scenarios and, from records, attempting to establish the probability of their occurrence and some measure of the consequences. From such a study it is possible to identify what action should be taken about each.

There are three possible decisions that may be taken about any risk:

1 *Ignore the risk.* This will be done if the probability of occurrence of the hazard, combined with the consequences of occurrence, is sufficiently low.
2 *Avoid the risk.* This can be done in many cases. For example, a town gas explosion in a block of flats like Ronan Point may be avoided simply by not connecting town gas to the building. Similarly, the possibility of a ship colliding with bridge pier can be avoided by designing a bridge that does not require a pier in the river. Whether this option is taken will depend on the probability of occurrence combined with the cost of avoidance and the consequences of occurrence.
3 *Design for the risk.* If the risk of occurrence is sufficiently high, then measures may be taken in the design to enable the structure to withstand the occurrence.

When a hazard cannot be avoided and is recognised as requiring action by the designer, then it moves from the unforeseen into the foreseen category. There are various ways in which the problem can be tackled.

If the hazard is likely to occur frequently, for example, if the structure is planned for a region where a hurricane was expected about once a year, then the structure would need to be designed to withstand the resulting loads without significant damage.

If the hazard is less frequent and/or the forces involved are much larger, design of the structure so that no significant damage occurs may be impractical. In such cases, the approach may be to design the structure so that it will remain standing under the action of the hazard, but in a damaged condition. This is the approach often taken for dealing with major earthquakes in seismic regions. Structures are designed to absorb the energy imparted by the earthquake by yield of some sections. These may be severely damaged but will not disintegrate. In this way, though repair costs may be very large, even requiring reconstruction, the risk of loss of life to the occupants can be minimised.

Fire is probably the commonest accident that can occur to a structure, and the approach taken to design for major fires is basically similar to that for earthquakes. The structure should remain standing under the action of the *design fire* for sufficient time for the building to be evacuated and for the lives of fire-fighters not to be put at unacceptable risk. Design for fire requires that, in a damaged state and when considerably lower load factors are used than those adopted for normal design, the structure will still be able to support its self-weight and some imposed load for a specified time. A standard fire is specified by a curve that gives the temperature as a function of time; however, it is also possible to define a design *fire load* for a structure and estimate a time–temperature curve from this. This approach can give advantages over the standard fire where it is known that the amount of combustible material expected to be present in the structure will be small.

Where does this leave the gas explosion in Ronan Point? The investigators who studied the collapse concluded that the pressure produced by the explosion was about 5 lb/in^2 which, in metric SI units, corresponds to 34 kN/m^2. Bearing in mind that the design load for a residential floor is 1.5 kN/m^2, it can be seen that this is a very large load and it would be unreasonable to design the structure to withstand it. Nevertheless, it was concluded that the probability of a gas explosion occurring in another multi-storey block of flats was such that the problem could not simply be ignored. The conclusion reached in relation to Ronan Point was that it was reasonable to expect that an explosion of the type that occurred would cause considerable damage in the region of the structure local to the explosion. What was not acceptable was that a relatively local accident should have led to such an extensive collapse. This led to the formulation of the principle now enshrined in our design codes that an accident or other unforeseen event should not cause damage that is disproportionate to the cause. Alternative ways of stating this concept are to say that structures should be designed so that 'house of cards' types of failure do not occur, or to say that structures should be capable of absorbing a substantial amount of damage before they collapse. A structure that will behave in this way is described as *robust*.

Common sense shows the main features that will lead to a robust design. The problem with a house of cards is that the removal of any card leads to the collapse of the whole structure. If the cards are fixed together, say with glue along the edges of the cards, then first it is clearly more difficult to remove a card and, second even if one card were removed, the structure would still remain standing. Thus a fundamental requirement of a robust structure is that the members are connected effectively and not just rested on each other. The problem with Ronan Point was that the connections between the members were too feeble to hold the structure together if one element was damaged or removed. A design approach was therefore proposed that members should be sufficiently interconnected that, if any single element was removed, the structure should remain standing. To avoid having to carry out the large number of calculations necessary to consider the stability of the structure with each member in turn removed, rules were developed for the required strength of the connections. In some structures, however, it is impossible to avoid situations where some key element cannot be removed without the whole structure collapsing. Consider, for example, a water tower supported on a single column. Failure of the column is bound to cause failure of the whole. These key elements, if they cannot be avoided, have to be designed to be extra strong.

It will be seen that the overall safety of structures is ensured in design by three levels of provision:

1   The loads that the structure is expected to support in service are designed for directly applying suitable safety factors to ensure that the probability of failure is very small and that, under the envisaged service conditions, no damage will occur.
2   The effects of accidental loads that have above a limited probability of occurrence, such as fire or earthquake in seismic regions, etc, are taken into account, but it is accepted that some damage may occur. Lower safety factors are applied to accidental loads.
3   The effects of events which are too improbable to take into account directly or are unknown, are covered to an acceptable degree by ensuring that the structure is robust and, though it may be damaged, the damage will not be disproportionate to the cause.

Building regulations and design codes

Building and the law – historical developments

In the previous section we discussed, at some length, how a uniform level of safety may be achieved in design. A question that was not addressed was: who decides what an appropriate level of safety should be?

In about 1800 BC Hammurabi, a Late Bronze Age king in Mesopotamia, finally combined the various small states of the area into a single state and set up his capital at Babylon. To ensure uniform justice over his dominions, he promulgated his 'Code'. This was a code of laws covering all aspects of life at the time. It is unlikely that his laws were original; they were almost certainly a codification and rationalisation of existing laws. It remains, however, the earliest written attempt to set out a uniform system of law. One provision of these laws deals with building. This can be translated in part to read:

> If a builder builds a house and he builds it not firm and it collapses and kills the owner of the house, then the builder shall be put to death.

This seems to be the earliest extant 'building regulation'; from Hammurabi's time onwards urban cultures seem to have found it necessary to control aspects of building by the use of law. There are good reasons why this happened, and these reasons have a relevance to the development of building regulations and structural design codes.

In general, and certainly until recently even in the UK, the quality of an item being offered for sale was not controlled by law. It was, and is, generally assumed that a purchaser can judge the quality of what is on offer and can decide whether the suggested price constitutes value for money. Unless the vendor offered some guarantee, the risk was the purchaser's (*caveat emptor*). In a free market, competition ensures that a balance between quality, cost and price is reached where the customer gets what they want at a reasonable price and the vendor's profit is not excessive. This process did not, however, work in the supply of safe housing because the purchaser was unable to judge whether what was being sold was, indeed, adequately safe.

As will have been seen from the previous section, safety, though apparently a simple concept, is, in fact, so complex that, even today, experts have considerable difficulty with it. In these circumstances, while the purchaser of a house could easily establish whether the house was big enough, whether it contained all the necessary features and whether it looked elegant, there was no way of judging whether or not it was adequately safe. The customer had to trust the word of the builder that it was. The builder, however, is aiming to make a profit and will try to cut costs to a level similar to, or below, that of the competition. In this circumstance, where safety levels cannot be judged by the purchaser and therefore do not come into an assessment of value for money, the builder will almost inevitably make economies in those aspects of the construction that exist to provide safety. Safety levels thus become depressed until the frequency of collapses become unacceptable to society and the law is used to remedy the situation. It seems reasonable to infer that this was the case in ancient Mesopotamia under Hammurabi, otherwise the law quoted above would not have been necessary and would not have been written.

In Rome, during the early empire, a similar situation existed and is documented. During the early empire, there was a major population increase in Rome and this put great pressure on the development of housing within the City. Not surprisingly, there was a tendency to build to increasing heights and this led to the development of the multi-storey tenement block. These could be six or more storeys high. Collapsing buildings became a relatively common occurrence and the emperors were obliged to promulgate decrees limiting their height.

Building law was not exclusively concerned with the problem of collapse and other aspects of safety related to buildings appeared in legislation from an early time. For example, Deuteronomy 22 Verse 8 gives the requirement:

> When thou buildest a new house, then thou shalt make a parapet for thy roof, that thou bring not blood upon thine house, if any man fall from thence.

In England in the Middle Ages, the main safety problem in towns was not the collapse of buildings but fire, and early building legislation was mainly concerned with limiting the possible spread of fire. The code of by-laws introduced by the first Mayor of London in 1189 included detailed provisions for fireproof party walls. It was, however, the Great Fire of 1666 that led to the first comprehensive set of building regulations. The Act for the Rebuilding of the City of London of 1667 provided very detailed prescriptions for construction, again mainly motivated by the requirement to prevent the spread of fire. All buildings were, in future, to be of brick or stone. Wall thicknesses, heights and the location of timbers relative to chimneys were all specified in detail. Further acts in 1707, 1708, 1724 and 1764 expanded these rules and made many of them more onerous. The Building Act of the 14th of George III was enacted in 1774. This set up the system of District Surveyors to ensure that the provisions of the Act were carried out by a system of approvals. The Act also further extended the rules for construction. This Act was considered to be excessively prescriptive by later generations and to provide a major block to innovation. Further acts were passed throughout the Victorian era that followed the same general principles. The highly prescriptive nature of these acts may be seen from the following extract from a London Building Act from as late as 1935.

> The external and party walls of a dwelling house shall for a wall of height of 50 feet, be of a thickness of $21^1/_2$ ins for the lowest storey, $17^1/_2$ ins for the next two storeys, and 13 ins for the remainder, when built of bricks not less than $8^1/_2$ ins long or of stone or other blocks of hard and incombustible substance, the beds or courses being horizontal.
>
> London Building (Amendment) Act 1935, HMSO, London

Though the prime motivation for legislation on building in England may have been the limitation of the spread of fire, it can be seen that they also provided for structural safety, since they specified the dimensions and materials to be used in the structure.

The significant point to note about these building regulations is their highly prescriptive nature. The builder was told, by law, exactly what to build and what dimensions to employ. In fact, in an age when the concept of calculating member sizes to resist assessed loadings had not been developed, it is difficult to see how any other approach to the assurance of safety could have been formulated if the approach used by Hammurabi was considered unacceptable. A fundamental problem with the system was that it totally stopped the introduction of any new form of construction or the use of new construction materials. Structural steel or reinforced concrete, both beginning to be used around the beginning of the twentieth century, could not be used in areas covered by the Acts, as the Acts did not include rules for their use.

The issue that has to be resolved by the employment of the building regulations in terms of structural safety is basically one of Quality Assurance. How can the public be assured that structures to which they have to trust their lives have been designed and constructed to appropriate standards of safety? The law may achieve this satisfactorily in circumstances where a fully-developed technology is being employed and where conditions have been constant over a long period, but it is not satisfactory in dealing with new, or changing, situations.

The professions as a means of control

An alternative approach for dealing with this problem had, in fact, developed during the nineteenth century. This approach is embodied in the concept of the Professional Chartered Engineer. To see how this alternative approach to maintaining standards, or quality, came about, a digression is required.

Groups with special skills have always been tempted to combine in some way in order to maintain a high price for their products, and circumvent the pressures of market forces. One manifestation of this in medieval Europe was the formation of craft guilds. These certainly existed in most large towns by the twelfth century and exercised monopolies within their town in their particular fields of activity. It seems highly likely that it became almost immediately apparent that prices could be maintained at the desired level only if the quality of the product was also controlled to a reasonably high level. If the quality slipped, the guild members would be put out of business by cheap imports from other towns. Guilds therefore aimed to ensure a uniformly high quality of product. They did this by instituting rigorous training and testing regimes for those wishing to join the guild and thus be allowed to practise the craft. This was achieved through apprenticeships. No doubt this also had the advantage of limiting the numbers skilled in the craft and hence further reinforced their ability to maintain prices.

At about the same time, other very different bodies were also finding it necessary to provide corporate controls on quality for competitive reasons. Again during the twelfth century, groups of scholars began to come together to attract students; a community of scholars being a more attractive proposition for students than lone scholars. Clearly, what attracted students was the quality of the teachers in the group and so the practice of licensing teachers following rigorous examination soon developed as a means of maintaining standards. This was the origin of organised universities in Europe, and the licences to teach were the origin of degrees.

Over centuries, this concept of self-governing bodies that controlled the terms of trade in a particular area, and also controlled training and entry to the body after rigorous examination of competence, became the model for the professions. Professions operated in rather different areas from the craft guilds, such as law and medicine. These were learned arts where the public for whom they provided a service were unlikely to be able to judge clearly the quality of what was on offer. This may appear like a licence to print money and in the short term it probably could be. In the longer term, however, a profession would only thrive if its practitioners were trusted and their superior expertise recognised, otherwise people would simply manage without their services. For example, you would not go to a doctor if you did not trust him and believe that your chances of a cure were substantially enhanced by doing so; you would simply carry on using traditional folk remedies. Similarly, if you were not convinced that a lawyer would be able to present your case better than you could yourself, you would not be prepared to pay one to represent you. The professions dealt with this situation by introducing, as well as the testing of the competence of candidates for entry, a code of ethics. Such codes typically required that members of the profession should put the interests of their clients ahead of their own interests and should not attempt to discredit the qualities of fellow practitioners in efforts to gain business. Advertising was also banned, so members of the profession had to rely on reputation among those who had used their services to recommend their excellence to new users. Members were also required to be individually responsible for the provision of a proper service and were not able to hide behind the protection of a limited company.

In its pure and ideal form, a profession has been defined by the American Society of Civil Engineers (and originally by Roscoe Pound of Harvard Law School) as 'a learned art practiced in a spirit of public service'. Unlike building, where the public was similarly unable to judge the quality of what was being offered, the quality of the service offered by the members of a profession was not controlled by law but by the profession itself acting as a qualifying and disciplinary body. When, early in the nineteenth century, engineering had aspirations to move beyond its craft origins, it set itself up as a profession and organised itself on exactly the model considered above. The civil engineers of the nineteenth century were not significantly involved in building (which was the province of builders), so the close definition in the building laws of what should be built did not affect them. They operated in areas such as railway construction, harbours and bridges, and the quality of what they did was assured by their training and by the professional institutions, not by law. In these areas, the professional system worked well for more than a century in ensuring adequate quality in the more 'high-tech', high-cost and high-risk areas of construction. It also permitted innovation on a large scale while, in building, the application of law almost completely blocked it.

It seems that the problem posed at the end of the nineteenth century by the introduction into buildings of new materials, such as structural steel and reinforced concrete, could have easily been solved by making engineers responsible for the design of buildings made from these materials and amending the building by-laws to permit this. This approach might have saved a great deal of argument within the civil and structural engineering professions over the last 25 years or so. Probably it is unrealistic to think that this might have been possible; the building industry could not have been expected to change suddenly and allow engineers into areas that had from time immemorial been the province of others.

The development of codes of practice

It should possibly not surprise us that the solution finally found to this problem of quality control in the design of buildings was a compromise. The case of reinforced concrete design will illustrate its development. Though originally invented in this country (the first patent for reinforced concrete was that of Wilkinson in 1854), its practical development was largely due to French entrepreneurs such as Hennebique. Hennebique and others, following successful developments in France, started to introduce reinforced concrete into the UK in the final years of the nineteenth century. These firms offered a design-and-build service. The design methods used were part of the stock in trade of the firms and were often kept secret. The result was that the client, or their representative, was unable to judge the quality of what was on offer or to compare the technical qualities of competing proposals. Furthermore, there was no basis for the inclusion of reinforced concrete into the building control system. The nature of the new materials made it unrealistic to try to control them by the type of prescriptive methods introduced into previous building acts and building by-laws. It quickly became apparent that the control would need to be applied to the design methods rather than to the dimensions and materials in the finished product. The solution almost universally adopted to solve this problem is the *code of practice* that attempts to define levels of safety and acceptable design methods for a particular type of construction. The nature and exact function of a code of practice has varied over time and also varies from country to country, depending on the methods of building control employed.

In the UK it was the Royal Institute of British Architects (RIBA) that was the first to try to develop a set of rules for the design of reinforced concrete structures. The RIBA set up a committee in 1905 and invited the Institution of Civil Engineers (ICE) to join. The ICE declined, which may have been short sighted but, as the discussion above indicates, the development of rules for design would have been against the basic professional concept. The ICE set up a rival committee to draft a 'state of the art' report to inform its members. The RIBA, it should be pointed out, was not primarily trying to develop rules to control the work of architects but rather to control the work of others whom the architect might employ. The RIBA published its rules in 1907. They were revised in 1911 and were incorporated into the London Building Acts in 1916, opening the way for the use of reinforced concrete within the capital. Other countries were working along similar lines at around this time. For example, French regulations were published in 1903 and American and continental cities, such as New York and Chicago and several German states, also developed sets of design rules that were included in their building regulations.

It should be noted that all the initiatives mentioned above were attempts to provide a means of controlling reinforced concrete design and construction. A very different initiative was, however, taking place in the United States where enthusiasts for the new material had set up the American Concrete Institute (ACI) in 1904. This body decided to draft its own code of practice for reinforced concrete design, the first version of which was published in 1925. The important difference between this document and the others was that it was produced, not by those wishing to control the use of reinforced concrete, but by the promoters of reinforced concrete who wished to assist designers in its proper and effective use. This code, and later versions of it, became very influential and was either adopted in total by cities or states or used as the basis of individual city regulations. The ACI is still probably the most influential organisation in the concrete field world-wide and produces revised versions of its code at regular intervals.

It would appear that by 1933, when the next major UK code of practice was produced (by the Reinforced Concrete Structures Committee of the Building Research Board of the Department of Scientific and Industrial Research), the approach had changed from a document intended primarily for control to one much more concerned with aiding and informing the designer. Experience with the ACI codes in the United States may have influenced this. This process has continued over the years and UK codes have increasingly become closer to design manuals, with the introduction of material such as tables of bending moment coefficients and even extensive sets of design charts. These are intro-duced entirely to assist designers and play little part in the function of the code as part of the building control system. The interaction between codes and build-ing control has also developed. Until the 1940s, building acts and bye-laws had design rules written into them (as mentioned earlier, the RIBA rules were writ-ten into the London Building Act of 1916, the 1935 code was written into the London Building Act of around that date).

From the introduction of the first post-war code (CP114:1948), codes were simply referred to by acts or by-laws as being satisfactory methods of design. More recently, local building acts and by-laws have been replaced by national Building Regulations. These simply state, in effect, that structures shall be safe. This require-ment is deemed to be met if design is carried out according to a document listed in a schedule of approved documents in the regulations. This list includes all relevant codes of practice. The local authority in the area where the building is to be con-structed has the responsibility of ensuring that structures are designed and built in such a way that the Regulations are satisfied. Design calculations have to be submit-ted to the authority, which has to check and approve them. If designers do not wish to follow a code of practice in the schedule of approved documents, then they are at liberty to use whatever method they wish, provided they can satisfy the local author-ity that their proposed methods are satisfactory and will lead to designs that are safe. In cases of disagreement between the designer and the local authority, the designer may appeal to the Building Regulations Division of the DETR. Modern UK codes are thus much more closely attuned to a 'professional' approach to structural design and are intended to provide guidance to properly-qualified designers rather than to provide a set of rules that must be obeyed.

In the UK, codes of practice are published by the British Standards Institution. They are written by a committee drawn from those elements of the construction industry and professions who have an interest in the particular technology being considered. The BSI provides a professional secretariat to the committee but the input from industry and the actual drafting effort is provided by the committee members who work on an entirely voluntary basis. Extensive discussions take place during the drafting process, both within the committee and with other experts and interested parties, to try to ensure that the resulting document truly represents a consensus of industry opinion. When a draft has been produced it is circulated to all interested parties within the industry for comment. Their comments are taken into account in the development of a final version of the code. The process of developing a code can take a long time; both the most recent versions of the design code for structural concrete (CP110:1972 and BS8110:1985) took about eight years to complete.

Limit state design

Codes set out rules for carrying out designs. It is helpful if these rules can be set out in a rational way. Providing a rational framework should aid understanding and should help to avoid mistakes in design. Such a framework was proposed in the CEB Recommendations for an International Code of Practice published in 1964. This is the *limit state* approach. The limit state method does no more than set out a logical framework for thinking about design and it has now been adopted by very many countries as a basic format for their codes. The new Eurocodes, for example, all use a limit state format. The basic concept is as follows.

There are many ways in which a structure can become unfit for its required purpose; for example, it may collapse, it may deflect too much or vibrate in such a way that it becomes unusable, and so on. In the limit state approach, the designer should identify all the ways by which the structure being designed might become unfit for its purpose, and ensure that each of these has been considered in the design. Each of these possible ways of becoming unfit is defined as a *limit state*. A structure, or element of a structure, is said to have entered a limit state if it has become unfit for its purpose. In order to check whether a limit state has been reached in the design, the following information must be provided. This may be defined by a code of practice or be arrived at by the designer from his knowledge of the purpose of the structure and the requirements of the client.

- Criteria defining the limits of acceptable behaviour. For example, the criterion for the limit state of collapse is that the structure shall be able to support its design load. Provided it can, then the structure has not entered the collapse limit state.
- Design loading under which the criteria should be checked.
- Design material properties which should be assumed in carrying out the check.
- A model of behaviour (generally a set of design equations) which should be assumed in order to calculate whether, when the design loading and material properties are assumed, the criteria are exceeded.

Limit states are commonly divided into two groups:

- *Ultimate limit states* are concerned with situations where the structure, or parts of the structure, are unable to support their design load and therefore fail.

- *Serviceability limit states* are concerned with issues such as excessive deflections. Such problems do not necessarily result in parts of the structure falling down, but still result in the structure being unable to fulfil its purpose satisfactorily.

Ultimate limit states are concerned with the structure not being strong enough, while serviceability limit states are generally concerned with the structure not being stiff enough.

Codes generally list the limit states that commonly need to be considered, define the design loads, material properties and criteria. Nevertheless, the designer should consider whether the limit states are all appropriate for the structure considered or whether there are further, special limit states that should be considered. The relevance of the criteria given in codes for serviceability limit states should especially be considered since these will tend to depend strongly on the precise purpose of the structure and the code limits may not be appropriate.

Debates about the form, purpose and content of codes

Despite the extensive, and conscientious, efforts made by code drafting committees to arrive at an acceptable consensus, it has to be said that full agreement on the form and content of codes is rarely achieved. As will soon be discovered by anyone who reads the correspondence column in the *New Civil Engineer*, or the Verulam column in the *Structural Engineer*, codes are a constant subject of heated debate within the civil and structural engineering profession. There are two basic types of issue which are debated.

The first class of subjects is the specific provisions in a code for a particular situation. Engineers have different ideas about how a particular element should be designed, and debate about such ideas is a fundamental step in increasing understanding and improving the provisions in future codes. Furthermore, code committees, like any other group of people, are fallible and occasionally make mistakes.

The second class of subjects debated is the question of what codes are for, what should be included in them and the basic form of the document. These issues generally produce the more heated debate, because codes are inevitably a compromise between a very wide range of views on these general issues. Furthermore, it is indisputable that codes have become longer and more complex with the passage of years, and this upsets many users. It may be helpful to the understanding of codes to try to summarise some of these views about the nature of codes. What follows are brief statements of, possibly, the more extreme positions, with some comments on each.

Codes should be short and simple; the current codes have become far too complex and difficult to assimilate compared with the older ones, which were perfectly satisfactory.

Codes have become more complex for a variety of reasons. First, we use higher-strength materials at much higher stress levels, and hence higher levels of strain than in the past. This has meant that the control of deflections and cracking has become much more important. Serviceability, which could largely be ignored with lower-strength materials, is thus becoming an increasingly important element in design. Pressures to economise in materials and to use longer spans with more slender members further increases the necessity to treat the design of members in greater detail. All this means that, although the old codes may have been satisfactory with the old materials and types of structure, they are not necessarily satisfactory in modern conditions. Second, in an increasingly litigious age, designers increasingly feel the need to obey codes more rigorously in the belief that, by sticking as closely as possible to an accepted consensus view of the appropriate method of design, they have a protection against claims of negligence, should anything go wrong. This introduces pressure on drafting committees to ensure that the coverage of the code is more comprehensive, since designers want provisions to deal with matters which, in the past, they would have been happy to handle on the basis of their own judgement. Third, research is constantly increasing our understanding of behaviour and enabling more realistic prediction methods to be formulated. Since the behaviour of structures is actually highly complex, more realistic methods tend to be more complex than previous highly-simplified models of behaviour. Despite these reasons, there is no benefit in having a code that is more complex than is absolutely necessary. Complexity leads to increased expense in design effort, increased likelihood of mistakes being made and, often, an illusory sense that the behaviour is being predicted with great accuracy. The problem is to pitch the level of comprehensiveness and complexity correctly. Whether it is or not is largely a matter of opinion, and hence the debate.

Codes should be recipe books providing a simple set of instructions which, if followed, will lead to satisfactory designs.

It is probably true that much design is concerned with simple repetitive structures that could be handled economically and adequately by a recipe-book approach. If it is possible to limit the scope of a code to these types of structure, then a recipe code could be produced. A number of attempts to do this have been made; in the concrete field, the Institution of Structural Engineers' 1985 Manual (the Green Book) attempts to do this fairly successfully. The production of recipe-book codes does not, however, cover the full scope and purpose of a proper code of practice, which is still needed as a base document and reference.

Codes should contain only the essential requirements (levels of safety, criteria, etc) and leave the methods of satisfying these requirements to the designer.

This is a very attractive idea in some respects, but in fact what most designers want from a code is a statement of acceptable design methods rather than a statement of the requirements. Most modern codes try to separate the requirements from the methods. In BS8110, for example, the safety format, partial safety factors, criteria and design material properties are all set out in a short initial chapter, while the remainder of the code provides the methods for satisfying these basic requirements. The new Eurocodes, which, when complete, will replace existing national codes throughout Europe, separate the material into Principles, which must be obeyed, and Rules, which contain design methods conforming with the Principles but where the user may use alternative methods, if they can be justified. It is, however, the view of many that, because design methods develop by experience and calibration with past satisfactory practice, the safety factors, criteria, material properties and design methods form a complete package that cannot properly be separated out. As an illustration of the problem of separating the methods from the requirements, it has been mentioned earlier that one of the factors covered by the partial safety factor applied to the material strength in limit state design (γ_m) is the accuracy of the design method. It follows that the value of γ_m depends on the particular design method adopted and that the two are therefore inextricably linked and cannot be considered in isolation.

Why do we need codes at all? Can we not get rid of them all and, as trained engineers, design using our knowledge and experience?

One answer to this has been given in the discussion of the reasons why codes were developed in the first place; they were necessary in order to allow new materials to be used within the existing framework of building control legislation. It could reasonably be argued that this could be changed and that the signature of an appropriately qualified engineer be considered an adequate guarantee of quality. While this has real attractions, it is contrary to the spirit of the age and most unlikely to be accepted. Furthermore, codes have real advantages for designers, which should not be lost sight of. The following may be mentioned in this context.

1   Codes ease communication between engineers, because designs carried out in accordance with a code follow a course that is well known to, and well understood by, other engineers. The format of the calculations is thus familiar.
2   Codes reduce argument, and thus save time and effort. Designs carried out correctly according to a code will generally be accepted without argument. It could take a great deal of time and effort to justify a design carried out on some other, unfamiliar, basis.
3   Codes form a readily accepted basis for the development of design software. In

fact, it is difficult to see how software houses could approach the development of commercial software if a document such as a code that sets out an agreed consensus did not exist.

4   By providing an agreed consensus, codes do provide the user with some protection against accusations of negligence, though this protection is not as cast iron as some designers would like to think; it still remains the responsibility of the designer to ensure that a method given in a code is appropriate for the particular problem being considered.

5   Codes provide a level playing field for the assessment of competing designs (this is one of the original reasons for developing codes and it remains true).

6   Codes provide an agreed basis for the drafting of textbooks and developing the content of degree courses. This is contentious and many would say that university courses should not be linked too closely with codes, but should aim to develop an in-depth understanding of structural and material behaviour. While accepting this aim, it should be noted that all courses contain structural design projects of various sorts and that these projects would be grossly artificial if they were not carried out using a code. Students will want textbooks that relate to this.

In summary, it seems likely that if codes did not exist then something very like them would have to be developed.

It must in fairness also be pointed out that codes have disadvantages. The most often-quoted disadvantage is that they tend to discourage innovation. This happens because the convenience of designing in accordance with a code means that the temptation to design outside the code is very slight, bearing in mind the commercial pressures under which designers currently work. As an example, suppose you wished to use some alternative design method for some aspect of your design. The new method may not take more time than the old one to carry out, though often it will, because computer programs will not exist for the new method and it will have to be carried out by hand. What will take time, however, is the development and presentation of a case to demonstrate to your colleagues, and to the local authority that has the duty of checking your design, that the new method is satisfactory. In all probability, you simply cannot afford this time and effort and so you will stick to the methods specified in the code. Thus, though codes in the UK are not mandatory, designers generally follow them fairly rigidly. Does this mean that codes have become prescriptive, and have therefore not achieved their purpose in removing the prescriptive nature of the old building acts?

Standards

There is a further set of documents that are important within the framework of building control and design activity. These are Standards. As industrialisation developed, it became essential for some elements of manufacture to be standardised so that parts became interchangeable. Possibly the earliest item to be formally standardised was screw threads. Nowadays, agreed standards to which articles or materials are manufactured are set out in British Standards. Like codes, these are produced by the BSI. In the construction field, almost all materials and products that might be used in a structure will claim to meet the requirements of the appropriate British Standard. By specifying that the materials should be in accordance with a British Standard, a designer can be sure of the quality and properties of the material to be used in the structure. Since uncertainty about the quality and properties of the materials used has a major influence on the values selected for the partial safety factors, the ability of a code to specify partial factors depends on material of reliable quality being supplied. Codes are therefore written on the assumption that the materials used in construction meet the requirements of the appropriate standards. The standards are thus an essential element in ensuring adequate safety.

Summary

It may be seen that there is a hierarchy of documents that form the total building control framework.

- At the top are the Building Regulations. These are produced by government and are part of the law of the land. They set out the basic requirements for buildings: in particular, that they should be safe.
- The next level of documents is the Codes of Practice. These give guidance to designers on methods of design that will ensure that construction will meet the Building Regulations in so far as this is within the control of the designer.
- Finally, there are the British Standards that specify the quality and properties of the materials that can confidently be assumed by the codes and by designers.

It will be seen that the development of codes arose from the requirements of building control and, initially, had no place in Civil Engineering design. However, the convenience of a system of codes has long been recognised in other fields and they have now appeared in all areas of construction. For example there are now codes for bridge design, the design of water-retaining structures, industrial chimneys and so on.

Eurocodes

Finally, fuller mention must be made of the *Eurocodes*. The development of the Eurocode system is the most ambitious attempt so far to develop a consistent, universal approach to structural design. The development is financed by the European Commission as part of its approach towards a Europe-wide system for the control of construction products, though the work is being carried out by the European Standards Organisation (CEN). As such, it is clearly intended as part of a system of control and the Commission has no mandate to develop documents to provide guidance or assistance to designers. This rather clearer function for the codes has an effect on the content and presentation of the documents. Currently there are nine principal documents in various stages of development. These are:

Eurocode 1	Basis of design and actions on structures
Eurocode 2	Design of concrete structures
Eurocode 3	Design of steel structures
Eurocode 4	Design of composite steel and concrete structures
Eurocode 5	Design of timber structures
Eurocode 6	Design of masonry structures
Eurocode 7	Geotechnical design
Eurocode 8	Design provisions for earthquake resistance of structures
Eurocode 9	Design of aluminium structures.

Each of these codes has a number of parts dealing with different types of structure. All the codes use the same basic partial safety factor, limit state format and so represent a fully consistent set of provisions. This is unlike most national sets of codes where the codes for different materials have been drafted by different committees with quite different ideas about how codes should be constructed. It has been mentioned that the development of a code for use purely within the UK can take eight years to carry through. Bearing in mind the difficulty of obtaining agreement among all the member countries of CEN, it should not be too surprising to learn that these codes have so far been under development for twenty years, and that none of them has yet appeared as a fully-agreed and operational code.

Within the European system of control, there are three levels of document that parallel those in the UK system. At the top is the Construction Products Directive, which is roughly equivalent to the UK Building Regulations, followed by the Eurocodes and the Euronorms, which are European standards.

Further reading

BS8110–1985. *The Structural Use of Concrete, Part 1: Code of Practice for Design and Construction*, British Standards Institution.

Comité Européen du Beton (1964). *Recommendations for an International Code of Practice for Reinforced Concrete*, American Concrete Institute/Cement and Concrete Association.

CP110–1972. *The Structural Use of Concrete, Part 1: Design, Materials and Workmanship*, British Standards Institution.

CP114:1969, *The Structural Use of Reinforced Concrete in Buildings*, British Standards Institution.

ENV 1991–1–1 *Eurocode 1: Basis of Design and Actions on Structures.*

Manual for the Design of Reinforced Concrete Building Structures (1985), Institution of Structural Engineers.

Mitchell, G. R. and Woodgate, R.W. (1971) *Floor Loadings in Office Buildings – The Results of a Survey*, Building Research Station Current Paper CP 3/71.

'Report of the Joint Committee on Reinforced Concrete' (1907) *RIBA Journal*, third series, XIV (15).

Report of the Reinforced Structures Committee of the Building Research Board (1933), HMSO July 1933.

Chapter 8

The design process

Introduction

The precise details of the tasks to be carried out by a designer (usually a consulting engineer) vary according to the nature and type of project. In a civil engineering project, such as a bridge or a dam, the engineer is likely to play a leading role, carrying out the design of all civil engineering elements, but will be required to co-ordinate the works of others and also to liaise closely with the client. In building projects it is usual for the structural engineer to be a member of the design team, which normally comprises architects, services consultants, quantity surveyors and a planning supervisor. Nowadays, a project manager is appointed to co-ordinate the work of the consultants and also acts as a link between the client and the team. Historically, the architect had this role, but fashions have changed.

The tasks of the engineer will also depend on the method of procurement of the project. In the traditional arrangement, the design team is retained by the client for the duration of the project to provide a 'full service'. This involves taking the brief and developing it, producing scheme proposals, preparation of tender documentation, providing detailed information to the contractor for construction and inspecting the works on site. A client who chooses a 'design and build' contract may require the design team to perform only limited services. For the engineer, typically this might be the preparation of performance specifications to enable tendering contractors to carry out the detailed designs. The engineer may or may not be retained to vet the contractor's designs. In some cases, the engineer may be novated to the successful tenderer so that the contractor becomes the engineers' 'client' for the detailed design stage. All these variants to the traditional method are meant to provide cost certainty to the clients. While an aspiring designer needs to become familiar with the various types of contracts, discussion here is concerned only with the design process. This is illustrated with reference to 'full service' in a traditional building contract. Although intelligent variations will be required to suit different circumstances, the basic principles should be obvious.

Factors governing the engineer's task

The final output of design is information in one form or another. Usually it involves drawings and specifications to the contractor and calculations to the Building Control Authority. It is important to remember that the client is concerned only with the end result and is not generally interested in the process of obtaining it. There are three main factors, that govern what and when the engineer provides information.

First, the contract between the engineer and the client will, either explicitly or implicitly, demand that the design work should be carried out such that:

- the structure is fit for its intended use
- the structure is durable
- the project is within budget
- the project is completed within an agreed programme.

These requirements imply that the engineer should produce sufficient and necessary information at the appropriate times.

Second, let us consider the requirements for price and programme certainty. The contractor is usually asked to price the project on the basis of a specification, typical drawings, geotechnical data for the site and a bill of quantities, produced on the basis of 'standard method of measurement'. The tender period will be relatively short, usually from four to eight weeks, depending on the complexity of the project. Tendering relies heavily on the accuracy of the documentation. Any variations (from the tender information) issued during the contract could give rise to a claim for increased costs or extension of time, or both. In short, the tender documentation should accurately reflect all matters that have an influence on price and programme. These will include:

- the type of construction
- the type of materials used
- quantities of materials used
- quality of materials (specification)
- operations involved
- special requirements/constraints affecting materials and/or method of construction
- conditions in and around the site
- performance requirements for items to be designed by the contractor, which will always include any temporary works necessary
- testing requirements.

In general the designer concentrates on 'what' to build, leaving the contractor to work out 'how to build'. The idea is to draw on the expertise of the contractor and hence derive direct cost benefits. Nevertheless, in complex designs, it will be helpful/necessary to produce a statement of any methods assumed in design.

The information to be provided should be obvious from the above. It should be accurate, lucid and sufficient if claims are to be avoided.

The last matter of concern to the designer is the requirements of *Construction Design Management* (CDM) Regulations. Under these, the designer is required to consider the health and safety of those engaged in the construction of the structure and of those engaged subsequently in cleaning and maintenance of the completed project. This entails qualitative consideration of the risks related to the materials specified and the construction process implied. Although not required to produce a risk analysis as part of the contract documentation, the designer is required to eliminate the risks at source where it is economically possible to do so. Where it is not possible to eliminate them, the designer is required to spell out the risks in a pre-tender Health and Safety Plan. It is reasonable for the designer to assume that the contractor is competent and fully familiar with all the normal hazards in construction. Thus only the exceptional risks need pointing out. The thinking here is that the contractor should be made aware of these at tender stage so that sufficient resources are built into the tender price to carry out the works safely. The contractor is entirely responsible for developing and executing the construction stage Health and Safety Plan.

Project stages

All projects proceed through stages, generally (but not always) as one continuous process. Typically, these could be briefing, outline proposals, scheme design, detailed design and tender documentation, production information, construction and post contract.

Briefing

This is a critical stage when the design team as a whole establishes the requirements of the client. In general, the brief needs developing jointly with the client. The initial ideas of the client may represent a 'wish list' and may not be consistent with the budget/programme. It is the responsibility of the design team to discuss the various alternatives for achieving the client's aim and the associated budget costs. The broad outline of a brief will emerge at the end of the scheme design stage. It must be written and circulated to all interested parties and must contain:

- the broad description of what is to be built, and where
- outline specification for the structure and all primary materials
- superimposed loading requirements

- operational parameters for the design services, e.g. temperature, humidity and lighting levels
- any constraints to be observed during construction and use
- any provisions to be made for future extensions
- budget cost and a programme.

During the subsequent stages of the project, the written brief must be revisited regularly and updated as the design evolves, always including the impact on costs.

Outline proposals

At the early stages of the project, only vague indicative sketches may be available from the architect. Nevertheless the engineer is required to provide preliminary information by means of advice, sketches, reports or outline specifications to enable the architect to produce outline proposals and the quantity surveyor to finalise an outline cost plan. At this stage, the engineer should visit the site, study and collate all relevant information about the site and the surroundings, contact the local authority to agree any matters of principle and also establish if there are any special precautions to be taken regarding ground conditions. The engineer should also advise the client on the need for geotechnical investigations and topographical surveys, and assist in the appointment of specialist contractors for these purposes.

Scheme design

During this stage, the outline proposals are further refined in the light of site findings and information from other consultants. At this point it is usual to consider alternative forms of construction/material. A set of scheme drawings should be prepared. It is also a good discipline to prepare a structural report to supplement the drawings. The report should:

- define the brief
- describe the alternatives considered
- illustrate the chosen proposals
- confirm the loading and other assumption made
- recommend further action where necessary.

This report will be valuable, especially if there is a delay in the approval of the scheme by the client and/or the planners. A programme for subsequent stages should also be agreed, subject to the client's approval of the scheme and budget.

Detailed design and tender documentation

As mentioned earlier, information contained in the tender documentation is a benchmark for the tender price and for assessing the effect of any subsequent variations. Effectively, the project is being built on paper and should, in theory, reveal all the problems the contractor will encounter on site. This will happen only if adequate drawings are prepared. At this stage the aim should be to provide the quantity surveyor with the same information that the contractor will require for construction.

With the above in mind, detailed design of the approved scheme should be developed in close liaison with the other members of the design team. Suitable calculations, drawings, estimates of reinforcement and final specifications for the works should be prepared. Engineers may be asked to comment on the list of contractors being considered for the project. The engineer should also advise on the need for any special conditions of contract. Information should be provided to the planning supervisor to enable him to prepare the pre-tender Health and Safety Plan.

Copy negatives of all the tender drawings should be made and marked as such, and no further amendments to these drawings should be made.

Production information stage

When the job is out to tender, final calculations and details should be prepared and the necessary information should be submitted to the Regulatory Authorities. It is also usual to prepare reinforcement drawings at this stage, bearing in mind the sequence in which the contractor is likely to require information. The need for a resident engineer should be agreed with the client, where appropriate.

Construction stage

During construction, the engineer is required to carry out intermittent inspections to guarantee that the works are being executed generally in accordance with the contract. There will also be formal site meetings to be attended. The frequency and stages at which the inspections are carried out is generally the prerogative of the engineer. Normally they would include all critical operations, such as early foundation works to agree the formation on which to found, typical superstructure elements and most reinforcement prior to concreting. In some jobs it may be necessary to witness site tests.

During this stage, the engineer should demand from the contractor method statements for critical operations before they are put in hand. The purpose behind this is to force the contractor to think about the likely problems in good time so that adequate precautions may be taken to avoid them. Also the engineer will have to comment on, and approve, contractor-designed elements and steel-fabrication drawings. Routine material testing, such as concrete cubes and bricks etc, should be undertaken, and the results should be carefully monitored and appropriate action taken in the event of non-compliance.

Post contract stage

In practice, during the contract, some changes are likely to have taken place in the light of site findings or errors or co-ordination with other disciplines. If so, the relevant drawings should be amended and annotated suitably.

The CDM regulations require the preparation of a Health and Safety file, which should be submitted to the client for safekeeping. The thinking behind these requirements is that the client will then be in possession of reliable information regarding the structure, which could be used in the future if alterations or repairs or maintenance are undertaken. With this in mind, the engineer should supply all 'as built' drawings, design loadings, geotechnical investigations, reports and a statement on any risks associated with routine operations (such as chasing walls for extending an electric circuit) for which engineers are unlikely to be consulted.

Concluding remarks

The above discussion should have highlighted the responsibility the engineer has at all stages of a project. Bearing in mind that the client is placing huge trust in the ability of the consultant to design and deliver the project, the duties must be undertaken with care and diligence. Appropriate information should be provided, at the right time, to minimise the risk of the budget being exceeded or the programme prolonged. However good a design might be, the client will not thank a consultant who cannot meet the needs of the budget and programme. This requires a pro-active approach to design and intelligent management of the design process.

Chapter 9

Designer in a changing world

Introduction

Many commentators believe that we are living through an age of profound technological change on the proportions of a second Industrial Revolution. Social, economic, environmental and political trends that we are witnessing now in the UK and Europe reinforce this belief. In the main, designers respond to clients' and society's needs and thus they need to be aware of the new challenges and opportunities that the new trends are likely to bring with them. A brief résumé of some of the issues relevant to the construction industry is given below.

Social change

There will be increasing demand for lifetime education, skills and training. Information and communication technologies will de-skill aspects of professionals' work and new business processes will require additional skills. The public will become more aggressive in its advocacy of some green and socially-desirable practices. There will be a move away from 'jobs for life' and most of us expect multiple jobs and careers.

Technological change

As a result of dramatic developments in the information and communication technologies, we might expect widespread use of simulations, modelling and virtual reality; increased use of robotics and computer-controlled automation to reduce human exposure to dangerous situations, electronic commerce, rapid growth in the volume of information and a growth in teleworking and homeworking.

Economic change

The enlargement of the European Union is likely to provide a ready supply of semi-skilled and unskilled workers. The less-educated members of society will find it increasingly difficult to secure full-time jobs on a permanent basis with good firms. Some major customers run global businesses and will be well-informed and more demanding of services and products. All businesses will need to work harder to maintain levels of growth and profitability, and will expect their assets to deliver more value, i.e. better quality and value for money.

Environmental

All governments throughout the world – the UK is taking a leading role – accept that there is a need to protect the environment and planet Earth. Global warming has been widely accepted as a fact of life. There will be continued increase in legislation, both from Brussels and Westminster, impacting on emissions to the atmosphere, energy use, recycling and water pollution. The planners are likely to introduce even tighter restrictions on land use such as development on brown field sites. Sustainability will be the byword.

This is the backdrop against which engineering design is practised. Successful designers will need to be multi-skilled, flexible and able to exploit modern communication and information technologies. They will also need to get things right first time to be profitable. Three particular topics, which have a bearing on these issues, are discussed in the rest of this chapter.

Continuing professional education/development (CPE/CPD)

Why CPE/CPD?

Engineers are professionals. They enter the profession on the basis of a reasonably accepted education, and undergo the necessary training before being recognised as chartered engineers. Enlightened individuals motivate themselves throughout their career so that the profession becomes an all-consuming way of life. There are many compelling reasons why continuing education and professional development are essential life-long activities for all engineers.

We are living in a knowledge-based economy. Engineering has always required knowledge and practice. The knowledge base itself has been expanding at a vast rate. New materials and techniques are being introduced all the time. A good professional has to keep up with the explosion of new information, to preserve a competitive edge in the market place. Thus enlightened self-interest can be a useful motivator.

A degree and chartered status cannot be ends in themselves. They simply lead to a place on the bottom rung of the profession ladder. It is simply not practical for the curriculum to cover every aspect of the field. An honest self-criticism and appraisal will reveal gaps in knowledge, and strengths and weaknesses in all individuals. These could be addressed as part of CPE/CPD.

From time to time, the profession is obliged to take on board radical and new approaches. Currently issues such as sustainability and the environment are likely to have a significant impact on our method of working. Then there are Eurocodes looming. All this will oblige engineers to tool up anew. This is nothing new; we have been through metrication, limit state design, disproportionate collapse, and so on. There will be further new areas in the future.

Information technology and computers are here to stay. Most clients expect engineers to embrace current technology in their work. Training is essential to exploit this fully. Clients also view critically the presentational skills of their professional advisers. While some of us are endowed with natural skills in this area, training and practice would help considerably.

In recent years all professions have faced considerable pressure from society to amend their traditional practices. Society appears to be sceptical of professionals and so it demands increased accountability. Thus there is a need to convince society that the profession is well-educated, trained and up to date. Enhanced public respect is essential to preserve the status of the profession.

Note that in the above discussion, no mention has been made of the compulsory/mandatory CPD requirements of the Professional Institution. This is deliberate. Enlightened self-interest rather than external compulsion would be a more sustainable reason for continuous self-improvement throughout anyone's career. Rewards accrue to individuals and to the community at large.

Types of CPE/CPD

The requirements and circumstances of individuals will be different. There cannot therefore be a uniform approach. As long as the individual remains motivated, there are many ways in which the aims discussed above can be pursued. It will be useful to set targets and goals for oneself and to review them frequently in the light of current progress. It is easy to excuse oneself, but critical self-examination and corrective actions are necessary if useful progress is to be made. Here are some thoughts.

It is important to cultivate the habit of regularly reading at least one quality national newspaper, technical journals, magazines and other publications in the allied fields. Good radio and television programmes also have a role in providing information on a broad front and also in identifying good and poor presentational methods. Clients are likely to be impressed by an intelligent and rounded individual.

Professional interaction is essential for self-improvement. This is most easily achieved in the work environment, provided individuals are prepared to discuss their problems openly with colleagues. Of course, it can be useful for overcoming a specific problem but can also be stimulating in other ways. Blind spots could be removed and new ways of approaching an issue could also be learned.

Participation in technical meetings of professional bodies will present further opportunities to interact with fellow professionals on a wider front. There is always something to learn from looking at other people's ideas and solutions to particular problems. This demands a child-like curiosity, which should also be nurtured by visits to construction sites and completed constructions.

Then there are a number of courses and conferences to attend. Clearly there is a limit to this, and one needs to be selective. Most employers will encourage such attendance, and these events bring benefits to the organisations as well as to individuals. It is important to write up the salient notes of any course attended, to reinforce understanding and assist future reference.

Postgraduate courses are available to practising engineers. Some are modular and allow the required number of modules to be built up over a period. There are others that require attendance on a day-release basis. There are also short courses on specific topics, such as fire engineering, soil mechanics and engineering dynamics. Typically, these last a week to ten days, spread over a period.

Another effective way to learn is through teaching, and through writing books. These demand research, re-visiting fundamentals and ordering random thoughts and ideas. Skills honed in this process will have a beneficial impact on professional work. Academic institutions will welcome industry participation whole-heartedly.

Participation in the affairs of professional institutions and trade bodies can be rewarding. It gives a different perspective on the industry and adds a new dimension to experience. Work in the field of standardisation also requires assistance from dedicated individuals. It is often thought of as a thankless task, but in truth it allows the profession to shape codes and standards to suit its needs.

In summary, there is no excuse for not pursuing CPE/CPD. A wide choice of methods is available. It requires dedication, some sacrifice and single mindness – all marks of a good professional.

Sustainability issues

The case for considering sustainability

As mentioned in the introduction to this chapter, sustainability is and will continue to be the byword in construction. The three main reasons usually given are that:

- we are currently using up the Earth's future resources
- we are destroying our own environment on a global scale in the pursuit of economic gain
- the resources of the Earth are not sufficient to sustain even the current human population of the world.

The Engineering Council in UK, in its publication *Guidelines on Environmental Issues* (1994) points out that:

- human activity currently uses 40 per cent of the biological products of the land, which in turn places severe pressures on the environment
- only 20 per cent of the Earth's population is responsible for this activity

- the remaining 80 per cent of the population aspires to the same levels of development as that 20 per cent, and this is simply not sustainable
- the world population is set to double from its current level in the next 40–50 years.

While experts differ about the existence of global warming, there appears to be reasonable evidence to suggest that doing nothing is not a luxury we can afford. The Meteorological Office in the UK has observed a mean global surface temperature rise of 0.6°C over the past 100 years. This should be seen as significant, as the average rise in global temperature since the last Ice Age (about 20 000 years ago) has been estimated to be only about 5°C. Also nine out of the ten warmest years have occurred during the last fifteen years. There is also increased glacial retreat.

What then is the possible reason for the recent increase in global temperature? The *greenhouse effect* is usually offered as an explanation. The Earth receives radiated heat from the sun. Part of this is absorbed by the Earth, and the remainder is reflected back into space. In the absence of an atmosphere around the Earth, the heat balance will result in a surface temperature of −18°C, which would be too cold to sustain life. Fortunately the Earth does have an atmosphere made up mainly of oxygen and nitrogen; but it also includes other gases including water vapour and carbon dioxide. These gases trap some of the heat, which would be otherwise emitted into space, and radiate it back to Earth. As a result the equilibrium temperature of the Earth's surface is about +15°C. This phenomenon is called the greenhouse effect and is basically essential for human comfort on Earth. However, the concern is over the increases in atmospheric concentrations of human-made carbon dioxide, methane, nitrous oxide and hydrocarbons. The dominant among these is carbon dioxide, which also has a very long lifetime (100 years) so that the effect of today's emissions will continue for another century. These increased concentrations are thought to be responsible for climate change and in particular global warming.

The many consequences of climate change, include:

- flooding, caused by sea level rises due to simple expansion of oceans and increased frequency of storms
- increased water stress (when more than 20 per cent of a country's annual water supply is used up)
- dying back of some tropical forests, which adds to the carbon burden
- crop yields increasing in high and mid latitudes, but decreasing in lower (tropical) latitudes, thus causing about 17 per cent increase in food prices
- effects on many aspects of human health, particularly the likely spread of malaria.

The above arguments have spurred many governments around the world to take action now and to consider seriously the sustainability aspects of our activities. The most quoted definition of sustainable development is from the Bruntland Report (1987) which states: 'Humanity has the ability to make development sustainable – to ensure that it meets the needs of the present without compromising the ability of future generations to meet their own needs.'

Engineering a sustainable development

Sustainability of the built environment requires a multi-disciplinary and a holistic approach to development. In theory a development must be able to continue indefinitely to be truly sustainable. A simple model, shown below, can be used to understand sustainability.

$$\text{energy} + \text{raw material} = \text{product} + \text{waste} + \text{pollution}$$

Note that this model also predicts waste and pollution as a linear function of input energy and raw materials, thus leading to the degradation of the environment.

Until now we have relied almost totally on resources that are finite and which are dug or drilled out of the ground. Our development course will eventually use up all the resources, however big the Earth is.

The challenge now is to move to a model in which we try to do our very best to achieve a closed loop in the use of energy and materials. In accepting that the resources are finite, this new approach will maximise the benefit obtained from materials and energy. Materials would be recycled and re-used in numerous different applications and, where feasible, energy would also be recovered. Emphasis will be more on quality of life rather than pure economic benefit alone. In reality we cannot achieve a completely closed loop. Sustainable development is about making the most judicious use of the resources we do have and make them last as long as possible in the hope that human ingenuity and technological advances can provide the long-term solutions to our problems.

A plan for sustainability will aim at:

- reduction of greenhouse gases
- more efficient use (and re-use) of resources
- minimisation of waste and constructive use of waste
- reduction of harmful effects on air, land and water, arising from building activities and occupation.

In particular, the following specific matters will be of relevance.

Environmental assessment

At the design stage of a project, a formal assessment of construction is required. This will involve the consideration of the use of particular materials and components in a building and subsequent operation of the structure and the method of eventual disposal at the end of the useful life of the building. The assessment will aim to minimise all environmental burdens. This is referred to as *life cycle assessment* (LCA). It is a 'cradle to grave' systems approach for understanding the environmental consequences of technological choices.

Life cycle costs (LCC)

The whole life cycle of projects should be understood and a true cost evaluated. In this context, *whole-life costing* is employed to measure the economic performance of investment. It adopts a building life cycle concept, starting with the installation and continuing over an agreed life for the building. It considers the initial cost, operational costs, replacement costs, residual value and discount rate. Life cycle costing is used to evaluate the best value for money when different alternatives are considered.

Embodied energy and operational energy

Energy is used in the extraction of raw materials for producing a product, the production itself, the transportation of the product to a building site and the installation into a structure. The sum total of this is referred to as the *embodied energy* of a material.

During the operation of a building or a structure, further energy is consumed for heating, ventilation, air conditioning and lighting, and by various types of equipment. This is referred to as the *operational energy*. Some studies have shown that the embodied energy does not vary significantly between different structural solutions provided they are based on sensible engineering. The studies have also shown that the embodied energy is modest in comparison with the operational energy. The ratio is about 10 per cent for a 60-year life. This suggests that focus should be on operational characteristics of the building. This in turn will lead to consideration of:

- orientation and shape of buildings
- insulation and air tightness
- ventilation
- comfort cooling/air conditioning
- careful balance between the provision of day-lighting and solar gain
- passive approach to using the structural fabric for energy reduction
- ice storage for mechanical options for cooling
- combined heat and power systems to eliminate the need for boiler and chiller plant in buildings.

Water usage

While the UK has had a plentiful water supply historically, the position is now changing. Water conservation and prudent use should be considered as part of a sustainable development. The measures could include:

- use of materials that are water-efficient in their manufacture
- incorporation of rainwater collection and storage
- introduction of grey water (derived from baths, showers and laundry facilities) recycling in large commercial developments
- specification of low-flush sanitary fittings and water-saving devices.

Development of contaminated land

Scarcity of land and planning restrictions will continue to force developments in contaminated lands. The contamination can be physical (old structures and obstructions) or chemical. Risk assessments based on study of the history of sites and thorough site investigations will be needed. Guidance is available to assess the risks of the levels of contamination and to undertake remedial action.

Concluding remarks

The world is at an environmental crossroads. Doing nothing is not an option. We have to change the habits and approach to construction drastically, to make the best use of our future resources. Our very survival depends on it.

Learning from failures

Introduction

It is the aim of every engineer to produce error-free designs. The basic approach to design involves considering failure scenarios and taking measures to avoid them. Thus, pro-active consideration of failure is part and parcel of all design. This pre-supposes that we can anticipate all modes of failure in all circumstances and, of course, this is not strictly true. It is because of this that many designers do not depart radically from well-proven constructions which have stood the test of time. However, innovation and the desire to do better cannot and should not be curbed. Engineers are often caught up in a dilemma between risk-taking and following a conservative approach.

In the field of civil and structural engineering, innovation is generally incremental and cautious. Nevertheless, in practice failures do occur. Possible general causes are discussed later. Critical appraisal of particular failures and their causes will be a valuable tool in the armoury of any designer. The development of civil and structural engineering is punctuated with major failures and these have had a seminal influence on the way we conduct particular aspects of design. These include the failures of the Tay Bridge in the UK (wind loading), the Tacoma Narrows Bridge in the USA (torsional vibration), the Westgate Bridge in Australia, and the Milford Haven Bridge in the UK (inadequacies in box girder designs), block of flats at Ronan Point in East London (progressive collapse of multi-storey construction) and the roof at Sir John Cass School in Hackney (high alumina cement components). Major events such as these culminate in public enquiries and, in some cases, lead to changes in regulations. All design offices will also be aware of un-publicised failures or near misses. It requires a sense of responsibility to evaluate these properly and adopt internal procedures to avoid their recurrence.

However dismal the impression from the above, the fact is that the construction industry is not in crisis, nor is it riddled with failures. Relative risks of death from various causes have been studied statistically and the fatal accident rate (FAR) of buildings falling down is several orders below many other risks commonly accepted by society, such as various modes of travel. Excessive obsession with failure also has an inhibiting psychological effect that will curb flair. Thus a careful balance needs to be struck between the illusion of risk-free design and a cavalier approach.

Causes of failure

Before considering the causes, it will be worth defining what is meant by 'failure'. It could include:

- total or partial collapse
- loss of safety, i.e. reduction in the factor of safety, without actual collapse
- reduction in load-bearing capacity without actual collapse, i.e. capacity less than the contractually agreed level
- loss of serviceability, e.g. sagging floors, excessive vibrations and cracking
- loss of durability, i.e. need for excessive premature repair well inside a reasonable design life.

While collapses are rare, all other types of failure noted above are common, particularly failures of serviceability and durability.

The likely causes of failure may be summarised:

- In the immediate aftermath of major failures the industry is usually very alert, but over the years complacency sets in. There is a tendency to court disaster through forgetfulness. In some of the allied industries, it is recognised that some mistakes are repeated every ten years. One possible safeguard against this collective amnesia is the organisation of frequent seminars and discussions entirely devoted to failures or near failures.
- Construction is a complex process involving a number of different parties, including the client. The number of interfaces increases the scope for things to

go wrong. It demands a disciplined communication that is both clear and timely.

- Gross errors in design and/or construction can lead to failures, which can be dramatic. For example, placing the reinforcement on the wrong face of a cantilever or the use of the wrong grade of bolts in steel connections, can have disastrous consequences.

- Feedback from insurers (who are concerned with claims that may not all be connected with failures) suggests some consistent leading causes over a long period. These are conceptual errors in design, inadequacies in supervision of both design and construction, and inadequacies in site investigations. While lack of knowledge will play a part, inability to allocate adequate resources for proper inspections and vetting of designs and details is self-inflicted. Irresponsible clients and their agents sometimes exploit difficult market conditions and pay fees far below the level necessary to provide a good service. This inevitably leads to short-cuts and possible failures.

- There are disturbing trends in the current methods of project planning. Allowing a reasonable time to develop the detailed designs is often seen as a luxury that cannot be afforded. Design is perceived as a mechanical process. The designers are thus unable to evaluate the interaction between various disciplines properly. This can give rise to failures in the terms defined at the beginning of this section.

In summary, the major causes of structural failures are:

- gross errors in design and/or construction
- inadequate supervision
- breakdown of communications.

Avoidance of failures

From the above analysis, it follows that not all failures can be avoided by increasing the safety margins in design. In fact, this plays only a small part. The relevant measures would appear to include:

- rigorous quality control procedures to ensure the use of appropriate materials
- adequate supervision on site by the contractor, and inspection by consultants, to ensure that the construction complies with the design intent
- technical audit of design
- good education and training of designers and construction operatives, so that the quality of the final product is at an acceptably high level.

In design, the following tips may be useful.

- One person should have overall responsibility.

- The basis of the design should be in writing.
- All changes to the design should be referred to the person responsible for the conceptual design.
- Lines and methods of communication should be clear to all.
- Each design is 'one-off' – beware of standard design.
- Beware of 'simple jobs'; they do not exist.
- Be extra vigilant in alteration work.
- Be super-critical of new products and methods.
- Do not let the computer rule you.

Caution with computers

It is now increasingly common to use computers for both design and drawing. Reliance on computers, without adequate safeguards, is fraught with problems. Designers are also regularly confronted with new software. It must be validated before it is accepted. The following points should be noted.

- All software contains errors.
- It is impractical to test a non-trivial program completely.
- Testing shows only the presence of errors *not* their absence.
- Reproducibility of the results of the standard tests of the program developer should be checked.

Even when using validated software, it is essential to take a rigorous and sceptical approach. Some of the strategies that could be adopted for verifying the results include:

- checking the output for obvious errors
- carrying out spot checks on the results (including 'back of envelope' calculations)
- checking the overall equilibrium
- checking that support restraints have been correctly applied by looking at deformations of the restrained nodes
- checking for symmetry (if present)
- checking the overall form of the results (Look at the deflected shape and the distribution of element forces/stresses. Do these agree with what was expected?)
- comparing results with those obtained from other (validated) programs.

Further reading

BSRIA Environmental Code of Practice for Buildings, BSRIA.

Core Objectives of The Institution of Civil Engineers, Institute of Civil Engineers.

CPD Requirements of The Institution of Structural Engineers, Institution of Structural Engineers.

Environmental Guidelines towards Sustainable Development of the Built Environment, Institution of Structural Engineers.

Environmental Issues in Construction, CIRIA SP93–94.

Guidelines on Environmental Issues (1994), The Engineering Council.

Indicators of Sustainable Development for the United Kingdom, Department of the Environment/HMSO.

Managing and Minimising Construction Waste, Institution of Civil Engineers.

'Opportunities for Change: a consultation paper on a UK strategy for sustainable construction,' (1980), HMSO.

Petroski, Henry (19XX), Design Paradigms, Cambridge University Press.

Reports of the British Government Panel on Sustainable Development, DETR/HMSO.

'Sustainable Development and Buildings' (1998), DETR Property Advisory Group.

The Reclaimed and Re-cycled Construction Materials Handbook, CIRIA C513.

Index

Page numbers in italic refer to illustrations.

abutments 10, 38
'acceptable modes of failure' 155
accidents 73, 79, 155–8
admixture 102
aggregate 100, 102, 103, 109
alkali silica reaction 108–9
alloy steel 113
American Concrete Institute 163–4
arches 37–9, *39, 58,* 125; and masonry 57, 58; problems with 60–1
Archimedes 13, *14*
Aspdin, Joseph 65

barriers 84–5, *84*
beams: analysis 13–30; continuous 44–7, *45, 52;* definition 10; deformation 27–9; design 26; failure of 19–20; height–depth ratio 34–5; load capacity 53–4; moments within 23–4, *24, 34;* sections 35–6, *36;* strain distribution in 27–9, *28;* stresses in 29–33, *see also* simply supported beam
'bearings' 74
bending 97, *97*
bending moment 36–7, *37, 66,* 76–7
bending moment diagram 20–4, *21, 24, 26,* 38; and continuous beam 44–7
bending strength 52
bins 72, 73
bitumen 58
blockwork 118
bolts and bolting 114, 128–9
bond strength 98
bowstring arch 37, *39*
bricks 56, 60, 118, *119,* 120; and fire resistance 99; and movement in masonry 124–5; properties 119
bridges 1, 5, 68, 162; cast iron 63, 139; and loading 77, 94
briefing 175–6
British Standards 169–70, *170;* for bricks 122, 123; for masonry 122–4, 125; for timbers 127–8, *see also* Codes of Practice
British Standards Institution 164–5
brittle fractures 63, 115
brittle materials 13, 33, 52, 65, 99
Brunel, Isambard Kingdom 140
Bruntland Report 184
buckling 6–7
Building Acts 160, 163, *see also* Codes of Practice
building industry 162
building regulations 158–71, 164, 170
building stones 57–9, 118, 120–1
buttresses:
and indirect actions 73–4

calcium silicate bricks 120, 123
cantilever 15, *15, 16,* 18, *18;* and bending moment diagram *21, 22, 23;* and moments 18–19; safety 152–3
carbon steel 106, 113; failure 49–53, *49*
carbonation:
of reinforced concrete 107
cast iron 63, 113, 139
cavity trays 123
cement 64–5, 100, 101–2, 121
ceramics 33
characteristc property of a material 143

characteristic value 79–80, 104, 142
chloride attack 107
cladding 66, 90–2, 119
clay 60, 119, 121
client requirements 2
climate: change 183; and loading 72
Codes of Practice 142, 143, 144, 149, 155, 170; and bridge loading 94; and building control 164; and concrete 108, 109, 111, 162–3; debated 166–70; development of 162–5; disadvantages of 169; and limit state approach 165; and loading 70, 79, 81; and steel 115; for timber 126; and wind flow 86–8, 89, 92
collision loading 94
Colosseum 65, 118
columns 10, 40–3, 42, see also struts
combination value 79–80
communication 168, 181, 188
composite materials see reinforced concrete
compression 40, 97
computers 181, 189
concrete 4, 64–8, 100–13; mix design 102–3; blocks 120, 121; brittleness 65; choice of 68–9; constituents 64–5, 100–3; fire performance 109–13, 112; flexural design 111–12; properties 64, 104–6; strength 103, 103, 104–6, 105, 106, 110; and stress–strain curve 98–9; surface reinforcement 111; and tensile strength 13; water content 103, see also various types of concrete e.g. reinforced concrete
construction stage 177
contaminated land 186
continuing professional education/development 180–2
continuous structures see structures, continuous
corrosion 116–17, 129
'corrugated iron' 64
costs 3, 185
courses, conferences and meetings 181–2
craft guilds 161
creep 67, 94, 99

damp proof courses 123
dams 68, 69
Darby, Abraham 62, 139
Dee bridge 63
deflection 5, 6–7, 43, 52, 53
deformation 11, 78, 98–9

design: acceptance of 168; assessment of 169; attributes 2–3; and changing world 180–90; definition 3; and information 174; loads and loading 70, 143, 166; process 173–8; software 168–9
design codes 158–71
design fire 157
design strength 143
detail design 176–7
deterioration mechanisms 100
dimensions of structures 137–8
disconnection failures 5–6, 43
District Surveyors 160
division by parts 90
domes 58, 125; of concrete 65; disadvantages 60
drag coefficient 90
ductility 52, 53, 99; definition 12–13
dynamic loads and analysis 77

earth: and bridges 94; as building material 68
earthquake loading 92–3
earthquakes 92, 156–7; and disconnection failures 6; and ductile materials 13; and dynamic loads 77; intensity measurement 92–3
economic change 179–80
Egyptian building 56, 58, 65, 118
elastic analysis 20, 53, 54, 140, see also plastic analysis
elastic instability 6–7
elastic modulus 12, 98, 105
elastic moment of resistance 31, 33
electric arc 114
embodied energy 185
environment 180, 182
equilibrium of forces 47
equilibrium of moments 47
Euler equation 153
Euler load 42
Euler's formula 42
Eurocodes 170–1, 180, see also Codes of Practice
European Standards Organisation 170

facing bricks 119
failure 186–9; avoidance 188–9; causes of 187–8; definition of 4–7, 187; disconnection 5–6; elastic instability 6–7; functional 5; of material overstress 7; of serviceability 5; translation 6; types of 4–7

fatal accident rate 187
fatigue 77, 78
fibre reinforced composites 13
fire 73, 77, 99, 160; design for 157; and
    material properties 78, 99–100, see also
    under paticular materials e.g. concrete
flanges 35
flexural strength 97
flooding 72
floor loadings 79, 80–1, 82, see also office
    loading
Forth Bridge 64
foundations 6, 10, 80
frames 63
frequency distribution 145–9, 147
frequent value 80
Freyssinet, Marie Eugène Léon 67
frost action 109, 122, 123
functional failure 5

Galileo 13–20, 15
galvanising 129
gantry girders 77
gas explosion 155, 157
glass 33, 51, 52; and elastic/plastic analysis
    53–4; failure 49–53, 49
global safety factor 150, 151, 152, 154, 155
global warming 183
glulam see timber, glued
Gothic period 59
gravel 64
gravity loads 70
Great Fire of London 160
Great Zimbabwe 57, 57
greenhouse effect and gases 183, 184
ground water 72

handrails 84–5, 84
Hankinson relationship 126–7
hardwood 127, 128
hazards 156–7
Health and Safety 175, 177, 178
heave of foundations 76–7, 76
heavyweight concrete 102
Hennebique, François 163
highway loading 94
hogging moment 44–6
'house of cards' failure 5–6, 43, 155, 157,
    see also progressive collapse
housing 60
human error 73
hydraulic cement 64, 65
hydrostatic loads 72

icing 72
indirect actions 73–7
Industrial Revolution 62–3, 139–40
information 174
information technology 181, 189
infrastructure facilities 2
innovation 169
inspection 177
Institution of Civil Engineers 163
iron: as construction material 62–4, see also
    various iron alloys e.g. wrought iron
Ironbridge 62, 62

joints and jointing: and masonry 124; and
    steel 114; and timber 62, 128–9
joist hangers 129

knowledge-based economy 180

Leaning Tower of Pisa 6
lever 13–19, 14
life cycle assessment 184–5
lime and limestone 58, 64–5, 108
limit state 78, 165–6
line of thrust 38, 40
litigation 167, 169
live loads 70–2
load 70–95, 81, 82, 141; bearing and
    transmission 8–10; capacity 53–4;
    deformation 98–9; design for 155;
    effects of 144; and safety 133–7, see also
    office loading
load deflection curve 50–2, 51
load intensity 11
load-elongation curve 98, 98
load-moment diagram 50–2, 51
loading models 94

masonry 6, 57, 68, 118–25; arches 38, 39,
    40; bonding 119; durability 123–4; fire
    performance of 125; movement in
    124–5; properties of 58; protective
    function of 56; strength 56, 118–19;
    thermal insulation 120; used materials
    in 119–22
material overstress failures 7
material properties 96–131; and limit states
    166
materials: and fire resistance 99–100;
    strength 26, 96–9, 137, 147; structural
    55–69
Mesopotamia 56, 158–9
method statement 175, 177

Milford Haven Bridge 187
Mitchell and Woodgate investigations 81, 135
modulus of elasticity 12, 98, 105
moisture 122–3, 124
moment-deflection response 49
moments 16–17, 35
mortar 57–8, 121, 122
movement 73–7, 74
'movement joints' 74
mud 55–7, 68

nails 128–9
negligence 167, 169
neutral axis 28–9
Newton's Third Law 19, 24

office loading 70, 71; and frequency
    distribution 81, 134, 135, 135, 136, 136
operational energy 185
optimisation 3
Ordinary Portland Cement 65

Pantheon, Rome 65, 102
partial safety factor 144–5, 148–9, 150, 168
partial safety factor method 142–5, 150, 152, 153, 154, 155
permanent actions 70
permissible stress 140–2, 150, 151–2, 153, 154, 155; and multiple materials 141–2
perpendicular style 59, 59
plastering 125
plastic analysis 53, 54, see also elastic analysis
plastic hinges 53, 54, 115
plastic moment of resistance 33
plastic rotation 50, 51, 52, 115
plasticity 99
plastics 33
plate steel 64
plywood 128
pollution 184
Pont du Gard 58
Portland cement 58, 101, 121
post contract stage 178
post-tensioning 67, 68, 68
pozzolan 64–5, 102, 109
pre-cast concrete 101, 102
pre-tensioning 67–8, 68
pressure distribution 89–92, 91
prestressed concrete 66–8
prestressing 66–7, 67, 68

probabilistic design 145–9, 146, 150, 152, 153, 154
probability of exceedence in one year 79
production information stage 177
professional bodies 161–2, 182
progressive collapse 6, 155
proposals in design 176
Pueblo Indians 56
purchases and the law 159

quality assurance 160–1
quarrying 57, 122
quasi-permanent values 80

railways 62, 63, 139–40, 162
rain 72
Rankine's formulae 42
reading habit 181
recipe book approach 167
recurrence interval 79
reinforced concrete 64–6, 100–13, 139;
    control of use 160, 162–5; deterioration
    66, 107–9; ductility 106–7; load
    capacity 151–2; materials used 106; and
    moment of resistance 34; and safety
    factors 141–2, see also concrete
relaxation 99
research 167, 182
return period 79, 135–7
Reynolds Number 90
risk analysis 156
roads 62
robustness 155–8
rolled steel 64
Roman architecture 58, 65, 102, 159
Ronan Point 6, 155, 156, 157, 187
roofs 62, 72, 82–4
rotational stiffness 48
Royal Institute of British Architects 163, 164
rubble masonry 57
rust 66, 116

safety 132–72, 175; and corrosion 138;
    definition 132–3, 145; and dimensions
    of structures 137–8; early developments
    139; and economics 132–3; factors 54,
    140; and fire conditions 99–100;
    formats 149–55, 154, 168; and house
    purchase 159; and loading 133–7; and
    material strength 133–7; and reinforced
    concrete 66; uncertainty 133–8
safety index 147–8, 148

sagging moment 45–6
St Louis Bridge 64
sandstone 122
scheme design 176
screws 128–9
Second World War 65
segmental bridges 68
seismic coefficient 93
serviceability 5, 167
serviceability limit state 78
settlement 76–7, 76
shear forces 24–6, 25, 26, 97
shear strength 97
sheet steel 64
shrinkage 75–6, 75, 94
silica fume 64–5, 102, 109
silos see tanks
simply supported beam 17–19, 21–2, 22, 23, 44; and bending moment diagram 21–2; and support moments 47–9, 48
slabs 9–10, see also beams
snow: for building 55, 56–7; and limit states 79; and loading 72, 80, 82–4, 83
social change 179
software for engineering 189
softwood 127, 128
soil 6, 10, 72
spalling 75, 110–11
standard deviation 147
statically determinate structures 49
statically indeterminate structures 49
steel: behaviour under loading 51, 52, 52; brittle fracture 115; and Building Acts 160; as construction material 62–4, 65–6; corrosion 116–17; and ductility 13; and elastic/plastic analysis 53–4; factors influencing choice 68–9; failure 49–53, 49; fire performance 99, 117–18; frame 8, 9, 63; and hardness 116; load capacity 150; plastic design 115; production methods 113–14; properties of 63, 115–17; in reinforced concrete 65; strength 115; stress–strain diagram 115, 116, 118; and temperature 110, 115, 117, 118, see also various types of steel e.g. carbon steel
Stephenson, Robert 140
stone and stonework 13, 57–9, 118, 120–1
storm damage 72
strain: definition 11; and indirect actions 73; and load deformation 98–9
strength of materials 26, 96–9, 142, 147
stress: definition 11; distribution 30, 32, 33; and fatigue failure 77; and indirect actions 73; and load deformation 98–9
stress–strain curve 12; of common materials 11–13, 29–33, 29; and deformation under load 98–9; for elastic-brittle material 29–31; for elastic-plastic material 31–4, 53–4; for steel 115, 116
structural form 55–69
structural steels: profiles 35–6, 36, 113, see also steel
structures: continuous 43–7, 45, 50, 50; definition 8; elements of 8–54, 11; historical description 13–22; testing of 140
struts 40–3, 42, 153; definition 10; deflection in 42–3, 43, see also column
sulphate attack 108, 121, 123–4
superimposed loading 70–2
superposition theorem 44
supports 94
suspension bridges 39, 41
suspension structures 43
sustainability 180, 182–6

tanks 72, 73
Tay Bridge 187
technological change 179
temperature 55, 56; and global warming 183; and indirect actions 73–6; and loading 72, 94; and material properties 78, 99, 109–10, 110; and movement of masonry 124, see also fire
tender documentation 174, 176–7
tensile strength 96–7
tension members 43
testing of structures 140
textbooks 169
thaumasite attack 108
theory of elasticity 20
theory of structures 140
thermal expansion 77
thermal insulation 55, see also temperature
tie 10
timber: bridges 60, 61; in construction 60–2, 125–30; durability 129–30; fire performance 130, 130; glued 128; and loading 126; materials used 127; properties 61; specification 127–8; and stress grading 126
timber-framed building 61–2, 61
time–temperature curve 99, 100, 100, 110, 157

traffic loading 94
translation failures 6
truss 10, 35–7, 36, 38, 61–2, 129
turbulence 85, 86, 89

ultimate limit states 78, 165, 166
ultimate stress 12
university courses 169, 182

vaults 58, 60, 85

wake 90, 91
walls 10
waste reduction 184
water 56, 72, 186
water tower 4

wattle and daub 56, 61
waves 77, 135
weathering steel 117
web 35
welding 77, 114
Wilkinson, John 65
wind 10; and direction 72; flow 89–92, 91; forces 55; gusts 85–7; and limit states 79; and loading 77, 80, 82–4, 85–92, 94, 135–6; speed 86, 87–9, 87, 88
wrought iron 63, 64, 113, 139

York Minster 59
Young's modulus see elastic modulus

zero plane displacement 89

Introduction to Design for Civil Engineers